INJECTION MOLDING TROUBLESHOOTING GUIDE, 3rd ED.

I0463108

The IM Troubleshooting Guide was originally prepared in 1996 as a 48 page convenient pocket sized resource for use in Injection Molding. This information is most useful by personnel who work in the injection molding field including press operators, technicians, engineers, etc. This 3rd ED is at 104 pages and includes selected extra pages from other APEBOOKS that are helpful in process set up and troubleshooting.

This book includes many useful definitions and tips for troubleshooting molding problems -- both process and tooling related. The book was written based on many years of process engineering. The solutions for correcting process problems are listed in the best order to solve the problem based on factors such as ease & timeliness to perform versus cost to implement and always considering effectiveness to solve problem. It is also useful to identify a common set of definitions for each department to use when discussing these common molding defects. Tips are often provided as to which defects may be process correctable versus those requiring product or mold changes.

An introduction to DOE and dimensional nominalization is made, but discussed in greater detail in some of the other booklets written by this author for injection molding ... these are listed later in this book ... a total of six books have been written for injection molding.

Copyright & ISBN

A major effort has been made to achieve a highly accurate and informative book; any comments regarding improvements via corrections or new material to be included in future printings is solicited and welcomed.

Re-formatted for epub and e-book distribution in 2011. Re sized to 5½ x 8½ in 2011 and pages added for the 3rd Edition to bring total page count to 104 pages.

Library of Congress, copyright number(s): TX0004354846; other copyrights apply to added pages ... TX0002998075, TX0004932677 & TX0005080970.

ISBN-13: 978-1466414341 .... ISBN-10: 1466414340

About the Author

The author - Jay Carender:
- Hands on processing skills,
- Mold build/mold design & engineering knowledge,
- Degree in Mechanical Engineering from GMI (General Motors Institute of Technology ... now known as Kettering University),
- DOE Training from Stat-Ease, Inc.,
- SPC Training from University of Tennessee Management Development Center,
- SPC Training from ASQC,
- Black Belt from AIT (Advanced Integrated Technologies),
- Productivity and Quality Improvement Training by Dr. Deming,
- Mold Design & Advanced Mold Design Training from New York University ... taught by John Klees Enterprises,
- Multiple courses on processing from RJG Industries, Inc.,
- Owner of Advanced Process Engineering - company started in 1990 to create and market pocket sized reference booklets for injection molding industry. Six booklets written & published along with other training manuals.

The hands on experience is from various fortune 500 companies performing injection molding. The extensive experience includes:
- Hot runner molding,
- High cavitation molds,
- High speed molding with cycle times less than 5 seconds,
- Stack molds, unscrewing molds, core pulls, slides, close tolerance parts,
- Engineering resins,
- SPC,
- Statistics,
- DOE to effect process improvement and dimensional nominalization.

Other APEBOOKS

Injection Molding Reference Guide, 4th Ed.
contains basic part design, trig tables, calc for thermal expansion w/ coeffs, SHCS data, torque specs, shrink data, cooling equation, mold debug guidelines, melt index data, resin density data, many tables of process guidelines, process development techniques, calculating heat load & water flow requirements, pipe data, conversion factors, transformer & motor current, PM & safety, basic statistics, equip selection guidelines and more.

Injection Molding Troubleshooting Guide, 3rd Ed.
contains troubleshooting tips/solutions for many injection molding defects, intro to DOE, discussion of VPT and Decoupled Molding^{SM} techniques (SM - RJG, Inc). This 3rd ED. includes many select pages from other APEBOOKS which are applicable to process set up and troubleshooting such as sources of variation and root cause analysis.

Math Skills for Injection Molding, 2nd Ed.
contains intro to basic algebra, using conversion factors, percentages, ratios, proportions, trig as needed for draft and tapers, trig tables, thermal expansion calculations, calculate shrinkage, determining part cost, understanding efficiency and utilization, intensification ratios and clamp tonnage, projected area, residence time, cooling time, interpolation, heat load, Cp, Cpk, Pp, Ppk, correlation, math equations and samples for calculating piezo and strain gage transducer full scale pressure, and more.

Pocket Injection Mold Engineering Standards
mold spec sheets, quoting & design direction, shrinkage, mold steels and hardness, heat treatment, thermal conductivity, thermal expansion, plating, surface finish tables, cooling design guidelines, gate designs, runner sizing, venting, sprue pullers, sucker pins, ejection, slides, support pillars, alignment guidelines, O-ring guidelines, hot runner info, torque specs, trig tables and more.

Managing Variation for Injection Molding, 2nd Ed.
understand & quantify variation, 6 sigma techniques, Cpk, Ppk, Z-score math, correlation, single & multi regression analysis to create predictive equations, DOEs, ANOVA, components of variance - how to quantify % each, MSE & Gage R&R, SQC, control charts, real time SPC, process mapping, process qualification & validation, FMEA, nominalization, molding techniques to reduce variation, and more.

Basic Statistics and SPC
basic statistics for operators, inspectors, set-ups, etc to prepare personnel for more effective SPC techniques, includes training on understanding SPC control charts (variable control charts - Xbar & R; Xbar & MR and attribute control charts - p, np, c, u charts ... how to compute control limits). Also included: what is variation, cause & effect diagrams, root cause analysis, histograms, pareto analysis and Gage R&R. Discusses inherent problems w/ X-bar & R charts as used in injection molding; explains better sub-grouping strategy.

The Advanced Process Engineering Guide
a compilation of the first five APEBOOKS in one book ... coming soon!

For questions or comments,
contact Jay Carender and Advanced Process Engineering at: advproeng@gmail.com

Table of Contents

Table of Contents (continued)

Note: Decoupled Molding^{SM} includes a Service Mark^{SM} by RJG, Inc.

Part Defects Defined

Blush (p. 7) .... Hazy or cloudy surface imperfection often near gate ... not splay or smear.

Brittleness (p. 8) .... Lack of strength, breaks easily when dropped or subjected to Gardner drop test.

Burns (p. 10) .... Usually a dark, burned area from dieseling at end of fill or vent location, but can appear elsewhere from overheated resin or in lesser cases of excessive fill speed.

Cloudy (p. 11) .... Lack of clarity in clear resins.

Deformed Part (p. 12).... Deformation caused by part ejection and/or mold opening (not warpage).

Dimensional (p. 13) .... Lack of dimensional compliance; too small or too large versus part print requirements.

Ejector marks (p. 17) .... Part deformation and/or stress whitening of part opposite side of wall from ejector pin.

Fish hooks & Flow Lines (p. 11) .... aka tails, J-hooks; surface imperfection looking like a fish hook or letter J ... a type of flow line, but different root cause.

Flash (p. 18) .... A thin layer of unwanted plastic at shut-off location, flash may be at parting line or at insert lines of mating mold components.

FM (p. 19).... Foreign material; may be contamination or proper material degraded.

Jetting (p. 20).... A twisting, widened flow line resembling the appearance of a snake or worm.

Mismatch (p. 21).... Incorrect match up or alignment of mating mold halves or inserted components.

Mold deposits (p. 22).... Surface defect caused by plastic or other residue left on mold's surface which results in poor molded surface.

Non-fill (p. 23).... Area that does not fill due to trapped air caused by poor or no venting.

Pulled .... Often used to describe deformed parts.

Short shot (p. 24) .... Incomplete parts (difference between non-fill and short shot relates to cause of each: mold problems such as poor venting or lack of wall thickness results in a "non-fill" whereas a short shot results from a process problem).

Sink (p. 25).... A dip or depression in part not following mold's surface; usually (but not always) adjacent or over an area of increased wall incorporating ribs, bosses or other part features causing increased wall thickness.

Smear .... Term used to describe splay or blush.

Specks (p. 19).... Dark, brown or black specks of burnt material or foreign material.

Splay (p. 26).... Cosmetic defect - silver streaks on surface.

Surface finish (p. 27).... Incorrect surface finish versus degree of polish, gloss or texture desired.

Void (p. 29).... An air pocket inside of the part wall.

Warpage (p. 28).... Dimensional problem; shape unlike that of mold; caused by non-uniform shrinkage.

Weld lines (p. 32) .... Lines or streaks in surface caused by two flow fronts coming together and joining.

Blush

Hazy surface imperfection often seen near the gate. Sometimes this term is incorrectly applied to splay, but blush can exist when there is no splay. Blush typically is located adjacent to the gate; thus, the term gate blush. Blush results from melt fracture as the material is forced through restrictive gates then expands into part wall.

Gate blush is most often caused by excessive injection speed for the resin grade and it's melt viscosity.

Best solution is:
1. Reduce injection fill rate -- reduce with velocity setpoints and not by reducing pressure. If reduced fill speed results in short shots, then profiled speed reductions may help (if machine is so equipped). Reduce the speed as melt front flows through gate; this may lay down a clean skin of resin without blush; can then attempt to increase speed as needed to accomplish proper mold fill. If mold is cold runner, maintain fast fill up to gate, then slow through gate, then experiment with the fast fill for remainder of part.

Other solutions include:
2. Increase the melt temperature.
3. Increase the mold temperature.
4. Increase gate size.
5. Check mold for coldwell; there should be a coldwell below sprue to collect the first slug of cold material in the nozzle.
6. Increase nozzle temperature.

Brittleness

Brittleness is often caused by a loss of molecular weight in the resin. Reduced molecular weight results in reduced physical properties -- reduced impact strength, reduced tensile strength and reduced elongation properties.

The best solution(s) to correct brittleness and avoid losses of molecular weight include (all three required):

1. Reduce melt temperature; use proper melt temps.
2. Process at reasonable barrel residence times.
3. Process hygroscopic resins with proper drying.

Do not rely on barrel setpoints as indicator of melt temperature. It is best to use a small diameter insertion probe pyrometer to identify the actual melt temperature (1/32 to 1/16 inch diameter results in low mass to heat). Many skilled Tech Service Representatives from resin suppliers use a rod with a ring on end to hold a paper cup. They use this to safely collect the melt sample and to isolate the melt from temperature losses. Follow your company's safety guidelines for this procedure; this should include appropriate gloves and faceshield. Collect the melt sample at a time when press is running normal cycles; interrupt cycle and immediately collect melt sample; probe or stir melt with melt probe. If shot is small in weight, you should preheat melt probe to approximate melt temperature before inserting into sample.

Please see the next page for more detailed discussion of residence time and calculation thereof.

Brittleness from degradation can also be caused by processing poorly dried resins (if resins are hygroscopic). A chemical reaction called "hydrolysis" occurs when such resins are melted in the presence of moisture. When hydrolysis occurs, a loss in molecular weight results which then causes a loss in physical properties.

Other solutions include:

4. Reduce nozzle temperature.
5. Resin may include too much regrind or contamination.
6. Reduce screw speed and/or back pressure.
7. Gate location affects orientation & directional strength.
8. Improve part design: wall thickness, ribs, etc.
9. Nylon parts absorb moisture from the atmosphere to increase toughness after molding (requires 24-48 hrs to reach equilibrium).
10. Use higher molecular weight resin.

Calculating Barrel Residence Time

When processing heat sensitive resins such as PVC, Polycarbonate, Polyesters, etc., it is useful to calculate the barrel residence time (time exposed to the elevated processing temperatures). Knowing this residence time is important for machine selection purposes or for troubleshooting degradation problems which might be caused thermally by excessive residence time. When residence times are calculated to be excessive, it may be necessary to locate an alternate molding machine or the process adjusted to make it more forgiving (i.e. reduce all or some barrel zone setpoints to reduce the melt temperature ... within limits of acceptable processability). Note: Hot runner systems will add some time to the total residence time; calculate hot runner system volume in ounces of styrene and add to the product of 1.4 times the barrel capacity.

Processing at the lowest possible process temperature and time will assist in maximizing physical properties. Maximum residence times of approximately 3 to 4 minutes are suggested for many resins (based on formula below ... it should be noted that this equation varies throughout the industry). The variation is typically the multiplier value 1.4 found at the beginning of the equation below. This multiplier is typically between 1 and 2 with its purpose to account for resin contained in the screw flights.

It should also be noted that a minimum residence time may also exist to properly melt crystalline resins. If the molding process results in very low residence times, then a hotter barrel profile may be required to properly prepare the resin for processing. Minimum residence times of 1.5 minutes are suggested.

Note: These minimum and maximum suggested times may vary for each resin; the resin supplier should be consulted for times and method of calculation.

The following formula is suggested. Minimum and maximum residence times with this formula should be per resin supplier guidelines or between 1.5 and 4 minutes if specific guidelines are not found. The example formula below would be for polycarbonate having a density of 1.20 grams/cu cc (substitute your resin's density in gr/cu cc where sp grav PC (1.2) is placed):

$$\frac{1.4 \text{ X bar cap (oz.)}}{\text{sp. grav PS (1.06)}} \text{x} \frac{\text{sp grav PC (1.2)}}{\text{shot wt (oz.)}} \text{x} \frac{\text{shot (oz.)}}{} \text{x} \frac{1\text{min}}{60 \text{ sec}} = \text{Time (min)}$$

This is constructed this way because molding machines are rated in polystyrene, the 1.4 multiplier is subject to debate and some use 2.0, but as long as you always use the same number, your answers will be relative to compare molding performance for a given residence time.

Burns

Burns are exactly that: burned and discolored material. You must first decide if the burn is from dieseling or excessive temperature. Dieseling results from trapping air in cavity -- typically last place to fill. The trapped air is compressed to point that it heats up and burns the plastic melt front -- this is same way a diesel engine operates. Diesel burns usually are evidenced by a black sooty surface burn whereas excessive temperature burns are deeper into the wall.

Solutions to reduced dieseling:
1. Reduce injection fill speed.
2. Clean all vents (also clean ejector pins since they provide some venting).
3. Reduce clamping pressure (in small increments; taking care to avoid flash).
4. Add vents (mold modification).
5. Change gate locations and/or number of gates.

Other burns are caused by excessive temperature or residence time (see also previous page on calculating residence time).

Solutions include:
1. Reduce barrel residence time.
2. Reduce melt temperature.
3. Eliminate contamination (other resins, etc.) if present.
4. Reduce fill speed (do first if dieseling).
5. Reduce back pressure.
6. Reduce screw rotation speed.
7. Improve material drying if hygroscopic or if condensation can occur.
8. Check T/C function and location; T/Cs should be preloaded into bottom of T/C well.
9. Check non-return valve on end of screw for resin traps or hold up locations.
10. Check for proper heater band function. A barrel zone with a burned out heater adjacent to it's T/C may result in a heater in the zone which overheats to satisfy the T/C and zone controller.
11. Operate at faster cycle times.

Anytime faster cycle time or reduced barrel residence time is suggested, a quick check of its efficacy is to interrupt cycle, quickly purge several air shots so that most resin in barrel has lesser heat soak time. Restart mold press cycle and gather shots to observe if and when defect returns. A defect which returns after 5-20 shots may be caused by excessive residence time or excessive melt temperature setpoints.

Cloudy or Hazy

Anytime a clear resin is cloudy, it usually is contamination. The contamination could be from the colorant from a previously run resin that hasn't been fully purged from screw or barrel. Contamination also comes from regrind -- grinder not fully cleaned, boxes, etc. Parts may appear cloudy due to poor mold finish -- lack of proper or desired mold polish. Some resins can appear hazy or cloudy if melt temperature is too low.

Solutions include:
1. Eliminate contamination in resin (especially regrind).
2. Purge barrel.
3. Empty and clean hopper and loader.
4. Check mold finish and correct as needed.
5. Increase melt temperature.
6. Increase fill rate/faster fill speed (may have packing with a cold melt resulting in highly stressed parts).
7. Increase mold temperature (may be better to reduce mold temperature for amorphous PET).
8. Reduce packing/hold pressure if in gate area.
9. Remove screw and clean check ring, screw and barrel.
10. Check and/or improve drying if resin is hygroscopic.

Fish Hooks and Flow Lines

This is a common problem when molding polycarbonate and acrylic, but can appear in other resins. Examine mold to verify that the lines are process related and not scratches in the mold. The flow lines are often shaped like fish hooks or the letters J or U. The lines often occur as the melt flows past ribs, engraving or other changes in mold steel surfaces.

Solutions include the following:
1. Decrease injection flow rate/speed.
2. Profile injection speed -- slow while melt lays skin over feature causing lines.
3. Increase melt temperature.
4. Increase injection packing and/or hold pressure.
5. Increase injection packing and/or hold time.
6. Increase mold half temperature.
7. Reduce depth of mold detail causing lines.
8. Add radius to mold feature causing lines.
9. Change gate location to alter fill pattern.

Deformed or Pulled

Part deformation may result from part sticking (partially or totally) in wrong mold half or during part ejection. You must first decide if problem is from mold opening (trying to stick in wrong mold half) or from ejection.

It is often useful to interrupt cycle after mold opens, but prior to part ejection. The parts should be tightly stuck to the ejector mold half with no signs of localized lifting. Any area that may be slightly lifted may have attempted to stick in opposite mold half. Another method to identify the mold location causing the problem is to lightly spray mold release onto mold where problem is suspected.

Solutions include:

1. Reduce packing and/or hold pressure. If this makes it worse, the problem may be excessive shrinkage onto cores located in stationary mold half (if sticking there) or excessive shrink onto cores in moving mold half if trouble occurs during ejection. If it is excessive shrinkage, then increased packing pressure and/or faster cycle times will help.
2. Reduce packing/hold pressure time.
3. Increase non-ejector mold half temperature.
4. Reduce ejector mold half temperature.
5. Check for equal length K.O. rods pushing ej. plate.
6. Slower mold opening speed (initial mold separation).
7. Slower ejection speeds.
8. Slower cycle times may help; if slower cycle helps it is best to continue with other items first before losing productivity. Slower cycles may help because of increased shrinkage onto cores in ejector mold half, but this might also be accomplished with reduced packing time and pressures also causing increased shrinkage.
9. Reduce mold undercuts.
10. Improve mold polish; eliminate EDM/other tool marks.
11. Increase draft in mold as needed.
12. Some resins have improved release from slightly abraded surface such as 600 - 1000 grit paper finish in direction of release (amorphous PET, K-resin, polypropylene, etc.) or a vapor honed finish.
13. Add drags to ejector mold half (adjacent to ejector pins). This is usually last resort as it is better to correct release problem from cavity. Add ejector pins as needed/if needed.
14. Add lubricant to resin.
15. Amorphous PET will stick to hot molds (over 150° F); thus, increased cooling may be needed if PET.

Dimensional Problems

The following does not include defects caused by warpage or deformed; see also warpage; see also deformed.

Dimensional problems are either process correctable or require mold steel changes. It is best to size the mold at an acceptable and planned process. Good engineering is required during process development to ensure that the development process can be duplicated in production (see later page in this book on process development press comparisons p. 65). Fill time during unit mold product development is often done too fast relative to the production press capability with a multi-cavity mold. At this planned process: the mold should be sampled, shrink calculated after 48 hours and mold dimensions sized to nominal print dimension plus shrinkage. If this has not taken place, then steel changes may be required if the mold is not process correctable.

If the mold has been qualified -- meaning it has run good dimensional quality at an accepted process then the following solutions can be tried to re-establish dimensional compliance. Dimensional checks should be after 24-48 hours when the part has fully stabilized.

PARTS ARE TOO SMALL / TOO MUCH SHRINKAGE
1. Increase packing pressure.
2. Increase fill rate (faster) to get less pressure drop.
3. Increase injection forward time; verify gate seal (procedure listed: Injection Molding Reference Guide, 4th Ed. & p. 66 this book)
4. Verify consistent cushion is maintained; if not adjust shot size and/or add back pressure. If still not maintaining cushion: check the screw's non-return valve.
5. Increase mold closed time: packing and/or cooling.
6. Reduce mold temperature; check for hot spots where adequate coolant flow may not exist.
7. Increase melt temperature (creates larger density shift which means more shrink, but reduced ΔP will yield better packing -- net result should be less shrink)
8. Enlarge the nozzle diameter and/or sprue size.
9. Reduce post mold ambient temperature.
10. Enlarge gate diameter or size ... if gates changed, then re-do gate seal study.
11. Use post mold cooling fixtures.
12. Add filler such as glass or talc (if acceptable solution).

PARTS ARE TOO LARGE / NOT ENOUGH SHRINKAGE
1. Decrease packing pressure.
2. Decrease cycle time by decreased cooling time.
3. Increase mold temperature.
4. Increase post mold storage temperature.
5. Experiment with adding a nucleating agent to some crystalline resins such as PP to increase crystallinity and resulting shrinkage.

Understanding Shrinkage

- Plastics are heated during injection molding
- Plastics expand when heated
- The total % expansion contracts during cooling and becomes the average shrinkage
- Shrinkage is altered by
  a. process packing pressures
  b. cooling rate (affected by water line location, water temperature, steel selection, area of steel versus plastic volume to be cooled, etc.)
  c. wall thickness (affects cooling rate)
  d. flow direction and retained orientation
- Shrinkage differentials may result in warpage
- Shrinkage is more predictable with amorphous resins and much less predictable with crystalline resins

Crystalline resins realize a greater amount of thermal expansion and contraction due to the shift from and to the orderly, closely spaced, crystalline structure versus the random amorphous structure. All resins are amorphous at melt temperature which is less space efficient relative to a crystalline configuration. This results in higher shrinkage for crystalline resins on the order of 0.018 in/in or 1.8% (approximately three times the shrink of amorphous resins). Fast cooling does result in some retention of the amorphous structure. Slower cooling results in a more crystalline structure and higher shrinkage. Keep in mind these cooling rate effects on crystallinity are minor; a crystalline resin will be predominately crystalline regardless of cooling rate. These, and other, process effects already mentioned make achieving close dimensional compliance with crystalline resins more difficult.

During the mold filling process there occurs a molecular alignment which can be partially retained during the cooling process. If the mold temperature is hot, then little retained orientation occurs. This is because the molecules will revert to the normal amorphous random structure, typical of all molten plastics. If the plastic cools without any retained orientation, then the molded part is said to be isotropic. Isotropic means the physical properties, including shrinkage, are the same in all directions. There typically is some cooling which takes place during and immediately after filling which results in retained orientation. This results in a part said to be anisotropic - meaning that the physical properties and shrinkage are not the same in all directions. A "living hinge" molded from polypropylene is a good example of a planned anisotropic condition; whereby, the molecules are highly oriented in the flow direction resulting in increased tensile strength in that same direction.

The shrinkage will be greater in the fill direction versus the direction 90° from fill direction (this characteristic applies primarily to unfilled crystalline resins).

Calculating Shrink & Cavity Sizing

Shrink rates are listed as inch/inch or as a percent ... one is same as other only two decimal places different (percent is "per hundred"; thus, the decimal point is two places to the right: 0.018 in/in equals 1.8 %). A polypropylene ruler that is 12 inches long with the aforementioned shrink rate would require a cavity length that is 12.220 inches long. This is derived from the formula below. Many people in the industry calculate this as follows: 12 X 1.018 = 12.216 inches ... this mold cavity is 0.004 inches too short, but who will notice. This math error continues today by some mold builders, but largely goes unnoticed because there is often a part tolerance that permits it and if the tolerance does not permit it, then the error is dismissed as actual shrink being different than predicted. Whether or not it is an error depends on method used to calculate the shrink.

Formulas for Calculating Shrink and Cavity Size

$$\text{Shrinkage} = \frac{(\text{cavity dim - part length})}{\text{cavity dimension}}$$

$$\text{Cavity Dim} = \frac{\text{finished part length}}{(1 - \text{shrink rate})}$$

Method #1 (correct method)

$$\text{cavity} = \frac{\text{finished part length}}{(1 - \text{shrink rate})} = \frac{12.000}{(1 - 0.018)}$$

$$\text{cavity} = \frac{12.000}{0.982}$$

$$\text{cavity} = 12.220 \text{ inches}$$

$$\text{shrink} = \frac{(\text{cavity dim - part length})}{\text{cavity dim}}$$

$$\text{shrink} = \frac{12.220 - 12.000}{12.220}$$

$$\text{shrink} = 0.018 \text{ in/in}$$

Note: The people who are said to calculate shrinkage incorrectly; would typically calculate shrink in the following manner which results in the CORRECT CAVITY SIZE IF AND <u>WHEN SHRINK RATE IS CALCULATED AS FOLLOWS</u>....the problem arises when a resin supplier's shrink rate is used and the supplier used different math to calculate the shrink rate. Typically on ultra critical dimensions, the mold will be left steel safe and tweaked or prototyped in a unit tool ... in either case correct answers can be obtained with either method so long as the right methods are used together.

Method #2 (alternate method that can work; see note above)

$$\text{shrink} = \frac{\text{cavity dim}}{\text{part dim}} = \frac{12.220}{12} = 1.01833$$

$$\text{cavity} = 12 \times 1.01833 = 12.220 \text{ inches}$$

Improving Nominalization

The nominalization can be improved by the following:

1. Optimal engineering in terms of predicting the actual plastic shrink rate[1].
2. Perform process corrections using the regression techniques learned in the DOE and regression pages found in the *Managing Variation for Injection Molding, 2nd ED* book (these are very powerful techniques for process correcting dimensional problems).
3. Number one above can be better achieved by adopting the following: measurements in different parts of mold should include a <u>balanced sampling of wall thicknesses and distance from gate</u>; this improves chances of including different cavity pressures and different cooling rates (cavity pressure will be reduced as you get farther from the gate & cooling rate depends on part geometry and mold design). Determine an average shrink rate for the X, Y & Z axis directions so that flow orientation is accounted for in the analysis (critical w/ high shrink resins like PE, PP, etc whereby the flow direction may shrink 30-40% more than crossflow direction).

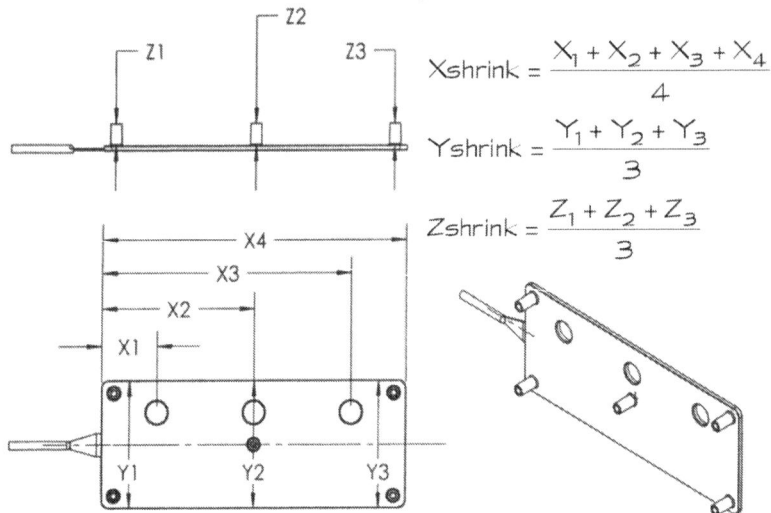

$$Xshrink = \frac{X_1 + X_2 + X_3 + X_4}{4}$$

$$Yshrink = \frac{Y_1 + Y_2 + Y_3}{3}$$

$$Zshrink = \frac{Z_1 + Z_2 + Z_3}{3}$$

5. Verify baseline process is good and typical for new mold (assumes that prototyping is being done via proto or unit tool).
6. Have same person measure both plastic and steel
7. Have both metrology dept and toolroom do measurements
8. <u>Reconcile the differences</u> -- note that if only one person measures the part or steel they are not wrong no matter what they get ... until the new mold arrives and the actual shrinkage is confirmed for each dimension; thus, take the time to perform parallel shrink determinations and reconcile the differences.
9. If a new mold is built w/o a proto tooling phase:
   a. design/build mold with CTFs as steelsafe w/ planned recut
   b. review similar mold in same resin, but still make CTFs steelsafe
   c. plan sufficient budget to recut or remake some components
10. The project engineer should be involved

---

[1] remember that every dimension may have a slightly different shrink rate; thus, the best average shrink rate must be determined ... deviations from average must be compared to the part tolerance.

Ejector Marks

Ejector marks are caused by local deformation of the plastic due to high ejection forces required. Sometimes the deformation includes stress whitening. The cause of ejector marks is typically one or more of the following:

1. Too much ejector force required by part for amount of ejection built into mold; add pins, etc. as needed or reduce ejector force required by improving draft or polish (less polish may be better with some resins; see p. 12, item 12, discussion of deformed or pulled part deformation).
2. Poor placement of ejector pins; not located directly below or adjacent to source of resistance -- whether it is an intentional drag/puller or normal part geometry; check drag/puller locations relative to ejector pins.
3. Ejection too soon while part is still too warm; slow cycle by increasing cooling time.
4. Ejection too late because part exhibits high shrinkage; run faster cycle so ejection takes place before part has time to shrink fully onto core.
5. Uneven ejection caused by knockout rods that are unequal lengths; correct K.O. rod length. Should use K.O. pattern that fully spreads K.O. forces over ejector plate (i.e. a 4" X 16" between centers pattern may be better than only a center K.O. or the 7" between centers pattern; 4 K.O.s are better than 2 or 1). K.O. rods should have lengths equal within 0.010" or better.
6. Part shape whereby a vacuum is created beneath part during ejection which greatly increases part retention onto core; in this case an air poppet is usually required to assist ejection with some air pressure; venting the ejector pins often does not solve this problem.
7. Long ejector pins making effective part wall thickness thinner than desired in one or more pin locations; correct ejector pin length.

Another type of ejector pin mark are those which are nicks received while part is free falling after ejection. The parts may hit the sharp edges of one or more ejector pins. These problems may not be easily solved, but the following are possible solutions:

1. Eject on the fly if machine is so equipped -- eject while clamp is still opening away from falling parts.
2. Reduce ejector forward dwell time if any.
3. Use a robot to unload parts eliminating free fall.
4. Try adding/using an air blow off.

Flash

Mold parting line shut-offs can be checked with PLASTIGAGE® clearance indicator. The green is PG-1 (aka SPG-1, MPG-1 & ZSPG-1, etc depending on supplier) for checking clearances of 0.001 to 0.003", the red is PR-1 for checking clearances of 0.002 to 0.006" ... there is also a blue (PB-1) for 0.004 - 0.009 and a yellow (PY-1) for 0.009 - 0.020 inch clearances. These wax strings can be laid across the parting line, then close mold with normal tonnage to check the actual clearance which may be causing flash. The wax strings are made such that the round string is compressed to a width based on clearance ... you compare the observed width with a scale included on packaging to determine clearance. This product can be purchased at suppliers to automotive engine rebuilders (auto parts store).

The best process solution to flash is to:
1. Reduce the injection pressure used for packing and/or filling. Sometimes fast filling, which likely uses high filling pressures to accomplish, will blow the mold open during filling. The mold may be flashed before the VP transfer to pack or hold takes place (VP -- from velocity control to pressure control); set pack and hold pressures to zero to determine if flashing takes place during filling. If during fill, reduce velocity setpoints which should reduce required filling pressure. If flash occurs during pack, then reduce pressures as needed.

Other solutions include:
2. Increase clamp tonnage if available. Less tonnage may help if mold is small (less than 2/3 tie bar space) resulting in possible platen wrap around the mold which creates clearance below mid-section of mold.
3. Adjust VP position and/or method; transfer earlier.
4. Reduce the shot size and/or cushion size.
5. Reduce melt temperature.
6. Improve drying of hygroscopic resins. Hygroscopic resins which are processed when wet can experience large drops in molecular weight to the point that the resin flows very easily. Moisture bubbles also can act like a plasticizer by reducing melt density which also makes the resin flow more easily.
7. Initial packing at a very low pressure to set up skin at location of flash, then increase pressure with next stage (hold pressure) to pack part as needed.
8. Reduce melt residence time if resin is degrading.
9. Check parallelism of machine's platens.
10. Check injection delay or pressure switch for injection; verify clamp tonnage achieved before start of injection.
11. Reduce pressure for restart after cycle interrupt.
12. Non-Process: Repair mold parting line/shut-offs; check mold design for adequate support pillars.

FM (Foreign Material); Contamination

FM may appear as brown or black specks. Regrind from decorated parts (painted, hot-stamped, etc.) can be a source of FM. Poor resin handling can also lead to FM or contamination. FM is typically used to describe contamination particles that are not miscible with the resin used; thus, it is very distinguishable versus contamination which may be partially or totally miscible with your resin, but may cause cloudiness -- see also cloudy. Avoid contamination by careful resin handling: both virgin and regrind. Use a drawer type hopper magnet and clean regularly; this may catch certain types of steel and iron.

Solutions may include:

1. Examine material handling system to determine how much of it contains the contamination if visible. Save a sample for possible analysis especially if found in the supplier's bags or boxes. Mark and segregate supplier bags and boxes for later disposition by supplier.
2. Purge barrel.
3. Clean the hopper, magnet, loader and material lines feeding press.
4. Switch to 100% virgin resin.
5. Check material lines for cracks and holes.
6. Clean grinders.
7. Check grinders for poor blade condition/adjustment.
8. Dedicate material lines for a specific resin type and/or color.
9. Clean mold vents and ejector pins.
10. Use Ionized air curtain if FM found to be on surface of parts; thus, post molding contamination. Ionized air curtains and blowers neutralize the static electricity charge that attracts dust and other contaminants.
11. Use a clean room to get clean room quality (surface contaminants).

Jetting

Jetting is sometimes called worm tracks as the flowing resin squirts into wall without laying down a normal skin. Normal flow results in a skin laid down like a tank or Caterpillar™ tread rolling across the ground with little or no shear at the wall surface. Melt flow from the point of resin entry (the gate) also fans out in all directions due to some amount of flow resistance. When there is little resistance because the gate is small compared to wall thickness, then jetting may occur as the resin enters and avoids contact with mold wall surfaces. Jetting most often occurs when there is this small gate versus the wall thickness AND the direction of gating results in no melt flow impingement onto core or cavity surfaces. A small sub-gate into a thick wall is not usually a problem as the flow enters at a 30°- 60° angle and impinges on core or cavity wall. Small tab gates are often in line with wall thickness and jetting may result if gate size is less than half of part wall. Filling rate also has an effect -- slowing the fill rate may make a poor mold design better.

Solutions include:

1. Reduce injection fill rate or speed. Profile filling speed so fill is slow while gate area is formed, then increase speed.
2. Increase melt temperature.
3. Increase mold temperature.
4. Increase gate depth. If mold has never run and jetting is a concern; plan gate depth to be at least 50% of wall thickness and prepare to go thicker after first sampling if needed. A gate depth of 60-80% of wall thickness is usually very safe.
5. Change gate location to accomplish some initial flow impingement on a mold surface.
6. If gate depth greater than 50% of wall is a problem; make gate wider as in a fan gate style.

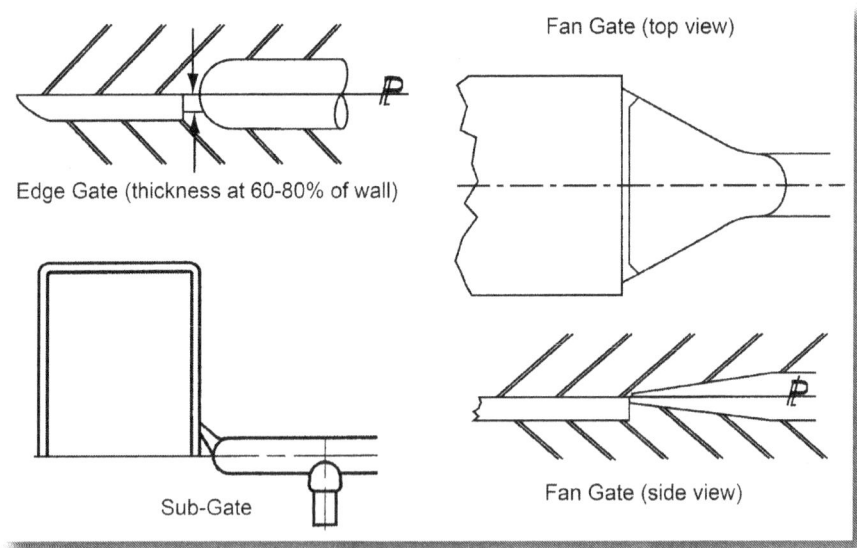

Edge Gate (thickness at 60-80% of wall)

Fan Gate (top view)

Sub-Gate

Fan Gate (side view)

Mismatch and Misalignment

Mismatch describes a situation having incorrect match or alignment of mating mold halves or inserted components. This condition is primarily a mold defect due to any of the first three items listed (item "D" should not be overlooked either); correction will typically require toolroom work:

A. Poor mold design and/or construction.
B. Mold wear such as worn leader pin bushings or locks.
C. Incorrect mold assembly (or reassembly).
D. Mold halves can slip in the press. If mold has slipped there may be a "thump" heard when mold closes as leader pin hits edge of bushing. Correct by: closing mold, support mold with overhead hoist for safety, loosen clamps (no tonnage), apply tonnage, then re-clamp mold to platens. An adequate number of mold clamps should be installed. The clamp bolt should be closer to the mold than the fulcrum point of the clamp (heel); mid-way is OK. If bolt is closer to fulcrum, then a longer clamp is suggested.

best if bolt is closer to clamp plate (mold) than the fulcrum pt.

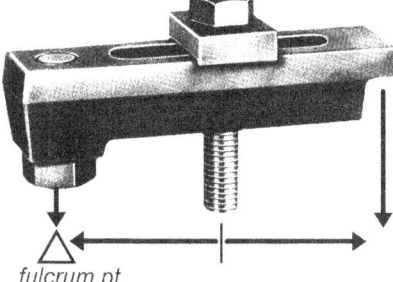

fulcrum pt

Some mold mismatch problems such as wall thickness differentials may be from poor core/cavity centering which may result from dirt being introduced to a pocket during re-assembly. A tall core having a chip or dirt speck can move the top of a core substantially, affecting wall thickness (item C. above).

Other mismatch considerations:

1. Proper mold construction and component positioning is always required.
2. Replace leader pin bushings as needed when worn. Do not rely on leader pins for very close alignment; use taper or straight (parallel) locks located at center lines of mold (12, 3, 6 and 9 o'clock positions). These positions are not affected by thermal expansion since the mold is located by it's center via the locating ring.
3. Take care to avoid dirt and chips during mold assembly. Designs with blind pockets for inserts may increase chances of such contaminants being introduced without detection.
4. Always cut molded parts from new mold tryouts and check wall thickness with micrometers (or pointed micrometers as needed). Correct mold problems when mold is new. Wall thickness differentials should not exceed 8% of maximum (for location). Min ÷ Max should not be less than 92% (may want to establish your own guidelines based on product requirements).

Mold Deposits

Some molded part surface defects are caused by mold deposits which consist of plastic or other residue left behind on mold's surface. A defect caused by the mold or mold deposit repeats itself exactly in subsequent shots. Deposits typically come from the resin additives such as mold release, colorants, anti-stats, etc. Deposits may also be very low molecular weight molecules of the base resin. Such low molecular weight additives and resins commonly exist in our molding resins and are volatilized in the heat of processing. When injecting air shots (purging barrel into air), observe volatiles given off of melt. These volatiles precipitate out onto the colder mold surface during normal molding. Some deposits come off with the ejected part or with subsequent molded parts; other deposits must be cleaned off with solvents and/or a light scouring action.

Solutions to reduce mold deposits include: We must first clean the mold using suitable method (without damage to mold's polish or texture). The following are tips to reduce or eliminate subsequent deposits from occurring.

1. Reduce melt temperature.
2. Reduce melt residence time in the barrel.
3. Reduce injection fill rate or speed. Fast fill rates create shear heating, especially in the gates.
4. Improve mold venting. Vacuum venting can lengthen the time interval between routine mold cleanings by pulling residue farther up vent relief or completely out of mold vent system.
5. Re-examine your resin's additive package and/or evaluate other similar resin grades which may be cleaner due to less short chain low molecular weight resin.
6. Check resin for contaminants.
7. Check molding surfaces for hot spots; some resins such as amorphous PET will stick to hot mold surfaces. A thin layer of the plastic may peel of the part and remain stuck to the mold steel like glue. A soft copper tool may be required to scrape this plastic from the mold. The hot mold condition will have to be corrected or additional deposits may form.
8. Establish mold maintenance cleaning schedule. Different mold types and resin molded influence this schedule, but a routine cleaning is often needed each 600 hours of molding run time.

Non-Fill

Sometimes used to describe a short shot, but more often used to describe an area that did not fill due to trapped air caused by poor venting. In the case of small details in the molded part, a short shot and a non-fill may describe the same defect. Many molding shops reserve the term non-fill for defects which are not process correctable; thus, requiring a part design change and/or a mold design change to correct the non-fill condition.

Solutions include:

1. Improve venting. Vents other than standard parting line vents may be required. Vents located in cores and cavities away from the parting line are harder to clean; thus, susceptible to loading with volatiles becoming ineffective. It is best to locate such vents in a moving pin which results in a somewhat self-cleaning action. Other vent designs might be pins that are quick change in design ... e.g. a pin that continues all the way to clamping plate where a removable access cover can be opened to remove pin for cleaning in the press (requires unbolting mold clamps from one mold half, then support both mold halves from opposite platen with the aid of overhead hoist support).
2. Change gate size and/or location.
3. Increase part wall thickness.
4. Change to resin having higher melt index (easier flowing).
5. Add flow leaders. Flow leaders are like a runner within the part wall; a groove or deepened channel may be added to the wall to deliver melt to the far side of a long flow length. Use caution when planning flow leaders with regards to affect on flow lines, weld lines and/or sinks where wall thickness is increased.
6. See also process solutions for short shots.

Short Shot

Short shots are under filled moldings. Sometimes "shorts" are easily seen since a part is missing a large portion/section of the part. Other times the short may be a hard-to-see tab or snap-fit latch detail. Experience will develop knowledge of where certain parts first begin to short when there is a process problem. Even well established and proven processes can result in short shots. The most common cause of shorts is the restart of a machine after a cycle interrupt (even a brief 10 second interrupt).

After a cycle interrupt, the mold will likely be colder than it is during normal uninterrupted cycles. A short period interrupt results in a colder mold versus normal equilibrium, but a longer interrupt often has the opposite affect whereby overheated resin (possibly degraded resin) is injected at normal process pressures resulting in flash (sometimes excessive).

Solutions include:

1. Allow process to stabilize after restart following a cycle interrupt. Many molders uses special conveyors or chutes to automatically discard 3-5 shots after the restart following an interrupt. Develop a more robust or more forgiving process if cycle interrupts are common. Develop procedure for operator to discard start-up shots after restart if automatic conveyors are not practical.
2. Increase packing pressure.
3. Increase fill rate or fill speed.
4. Increase mold temperature.
5. Increase melt temperature.
6. Improve mold venting (see section on non-fills).
7. Check process for presence of a consistent cushion. Using zero back pressure may result in inconsistent shot plastication. A faulty non-return valve on screw tip may also result in inconsistent cushions and shorts.
8. Adjust transfer point and/or transfer method ... e.g. an early transfer at 80-90% of fill may result in shorts if the next stage is too slow of velocity or too low of pressure; adjust transfer to 95% fill.
9. Increase gate and/or runner size.
10. Check gates and nozzle for blockages.
11. Flow length may be too long: reposition gates, increase part wall thickness, add flow leader to part, add gates or change to resin having higher melt index (easier flowing).

Sink

A sink is a dip or depression in the part which does not follow the mold's surface; usually (but not always) adjacent or over an area of increased wall incorporating ribs, bosses or other part features causing increased wall thickness.

Solutions include (nearly the same as for voids except for #3 - mold temperature):

1. Increase pack and/or hold pressures.
2. Increase injection forward time.
3. Reduce the mold temperature. Reducing both halves may work fine, but reducing the side with the most visible sink (most visible side for part as assembled or in use), and increasing the other side may permit longer packing while freezing the visible skin (experiment with mold temperatures).
4. Adjust injection speed (typically reduced injection speed, but maybe faster injection speed).
5. Reduce melt temperature.
6. Increase gate and/or runner size. An undersized gate will result in premature gate seal before sufficient packing has occurred. Verify injection forward time does accomplish a gate seal. Full round runners have least pressure drop and remain fluid for packing longer; trapezoidal is next best; half round runners are not recommended.
7. Verify that a consistent cushion is accomplished. If not, adjust shot size and make sure there is some back pressure for good consistent plastication. If still not a consistent cushion, check non-return valve on end of screw (done by maintenance department).
8. Increase nozzle diameter if possible; check for nozzle obstructions -- remove as needed.
9. Some parts may permit design changes allowing material saver steel inserts or pins to be used to core out material beneath thick sections.
10. It is always best to design part and or mold so that fill is to a uniform wall thickness or at least from thick to thin. Thin to thick will often result in sinks or voids.
11. Make interior ribs or wall intersections at only 50-60% of wall thickness for visible wall surfaces.
12. Thick walled parts may be ejected with a hot molten center of wall thickness. This heat may result in voids or sinks; experiment with post mold cooling: air versus water and different water temperatures.

Splay or Silver Streaks

Splay and silver streaks are often the result of processing hygroscopic resins which have not been sufficiently dried. Other times the appearance may be from excessive heat which burns off volatiles and/or moisture that may be present (even after proper drying). This excessive heat may come from barrel setpoints, too much back pressure, too fast of screw speed or too fast of injection speed causing shear heating as resin passes thru restrictive gates.

Solutions include:

1. Check dryer for proper operation; also check for proper drying residence time. All resins brought indoors during winter months are susceptible to surface condensation even when dewpoints are low.
2. Check barrel setpoints and actual melt temperature to be within supplier guidelines. Reduce barrel setpoints.
3. Reduce injection speed.
4. Reduce screw rotation speed.
5. Reduce back pressure (but maintain some back pressure).
6. Reduce nozzle temperature.
7. Reduce or eliminate melt decompression (suckback). Splay can occur with resins that are not hygroscopic by the introduction of moist ambient air during screw decompression.
8. Change to larger nozzle orifice diameter (but smaller than sprue orifice diameter).
9. Check mold cores and cavities for condensation. Summer dew points are often near 75° F; thus, if your coolant water temperature is below 75° F, you could have small, hard to detect levels of condensation. Increase coolant temperature. Reduce mold open time and/or cycle time so that actual mold core/cavity temperatures do not fall below dewpoint. Air condition (dehumidify) the mold room air -- local to mold or entire room.
10. Check mold for water leaks. Seeping O-rings may introduce small amount of moisture to cause splay. Observe sitting mold for water droplets at insert lines.
11. Clean mold parting line including vents.
12. Check resin for contamination including excessive "fines". Fines -- resin dust -- have a high surface area relative to it's mass; thus, acting like super sponges in their ability to quickly absorb ambient moisture.

Incorrect Surface Finish

An incorrect surface finish may be the result of an incorrect specification of mold finish or inadequate mold polishing. See also the *Injection Molding Reference Guide, 4th Ed.* which discusses various SPI designations of mold polish. Incorrect surface finish may also result from mold damage whereby the mold's surface has changed from erosion of the mold steel. Mold steel erosion is common with glass filled resins, but other resins can also be abrasive. Mold texture opposite a gate can easily be worn away if not protected by high hardness surface treatments such as Titanium Nitride. Erosion can also be caused by dieseling at areas of poor venting. See also mold deposits as a possible cause of incorrect surface finish.

Solutions to achieve proper mold finish include:

1. Verify proper steel finish in mold.
2. Check mold for mold deposits.
3. Adjust mold temperature (up or down). In most cases a warmer mold will help the resin duplicate the actual steel finish. This is especially true with many fine EDM finishes where a warm or hot mold is required for resin to properly fill the small "pits" of the surface. Very hot molds can be used with glass filled resins to get resin to wet out a polished mold surface between steel and glass fibers.
4. Increase melt temperature.
5. Increase packing pressure.
6. Increase injection fill rate or speed.
7. Keep all steel tools out of mold. Soft copper tools and picks are best for re-moving stuck moldings. Certain grades of brass will work with high hardness tool steels, but could scratch highly polished P-20.
8. Post mold handling of parts molded from fine EDM textures is also critical in that the molded "peaks" of the textured part surface are easily scratched off leaving lines. For this reason, a fine EDM finish may not be advisable with certain resins.

Warpage

Warpage can be a difficult problem to overcome due to complex part shapes and ability to accomplish equal cooling on both sides. Warpage primarily results from shrink differentials in the part. Shrink differentials result from uneven cooling which may come from part design or mold design. Filling patterns which affect molecular orientation also result in shrinkage differentials.

Solutions include:

1. Adjust mold half temperatures. When attempting to process correct warpage, it is usually necessary to have separate coolant temperature control between mold halves. Plastic often will wrap concave toward the hotter mold surface with all else being equal. Check water flow in all circuits. If coolant temperature differentials are needed, the cycle will be favorably impacted by cooling the hot surface down rather than adding more heat to the other mold half. NOTE: The same water coolant temperatures may result in different actual steel temperatures due to the mold design; correct with mold design changes or aforementioned coolant temperature differentials. Corners of square shaped parts typically have more cooling in outside of wall (cavity) than the inside wall (core); this is common unless special cooling considerations are applied by the mold designer.
2. Increase or reduce injection packing pressure. Rectangular and Elliptical shaped parts can have the short and long axis greatly affected by fill speed and packing pressure; thus, experiment with combinations of each (often depends on gate location).
3. Adjust fill rate or speed (typically faster is better).
4. Move gate to change location of retained orientation.
5. Design part to have more uniform wall thickness with gradual changes in wall thickness if required.
6. Reduce injection forward time.
7. Increase cooling time.
8. Verify presence of proper cushion and gate seal.
9. Modify part design so that longitudinal ribs or long sections of thickened wall are interrupted.
10. Use a resin with a lesser shrink rate; molded part shrink differentials will be less resulting in less shrinkage in a given mold or part design.
11. Increase gate size to have less orientation.
12. Check for sufficient ejectors.

Voids

Voids are often ignored if not visible, but some molders supplying very critical components must cut open parts on regular basis to detect voids or they may x-ray the parts looking for voids. Voids can cause a significant loss of strength if not corrected. Solutions include:

1. Increase pack and/or hold pressures.
2. Increase packing or hold pressure time (injection forward time).
3. Raise the mold temperature, especially with thick walled parts. If a colder mold is used, then the outer skin formed by core and cavity surfaces can be frozen or fixed which results in the plastic shrinking more from the inside resulting in a void. By increasing mold temperature, the cooling rate of the skin is slowed so that entire wall shrinks more uniformly resulting in lesser or zero voids. Note: The warmer molds will likely increase cycle. Significant jumps in mold temperature may be required (e.g. +40-50° F).
4. Adjust injection speed (typically reduced injection speed).
5. Reduce melt temperature.
6. Increase gate and/or runner size. An undersized gate will result in premature gate seal before sufficient packing has occurred. Verify injection forward time does accomplish a gate seal. Full round runners have least pressure drop and remain fluid for packing longer; trapezoidal is next best; half round runners are not recommended ... see also next two pages on runners and runner sizing.
7. Verify that a consistent cushion is accomplished. If not, adjust shot size and make sure there is some back pressure for good consistent plastication. If still not a consistent cushion, check non-return valve on end of screw (done by maintenance department).
8. Increase nozzle diameter if possible; check for nozzle obstructions -- remove as needed.
9. Some voids may be from moisture; thus, proper resin drying may be required.
10. Some parts may permit design changes allowing material saver steel inserts or pins to be used to core out thick sections.
11. It is always best to design part and or mold so that fill is to a uniform wall thickness or at least from thick to thin. Thin to thick will often result in sinks or voids.

Runners

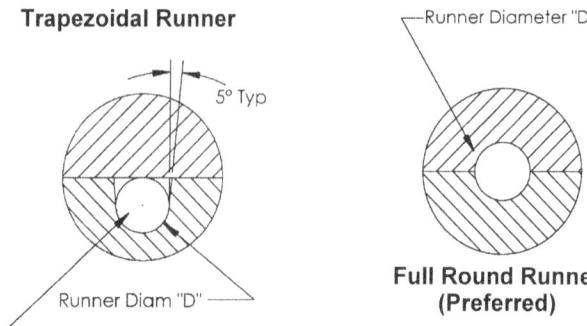

Trapezoidal Runner

5° Typ

Runner Diam "D"

Runner Diameter "D"

**Full Round Runner
(Preferred)**

This is typical design for trapezoidal runner. A full round runner is best, but requires cutting runner in both mold halves. Not usually a problem, except when stripper plate is planned for ejecting runner such as three plate molds.

What makes full round best, and this design for trapezoid next best choice, is the ratio of CSA (cross sectional area) versus surface area. To accomplish lowest pressure drop we need largest CSA versus surface area. The surface area is where the heat transfer (cooling) will take place. We do not want premature cooling of the runner while fill and pack take place. Ideally there would be no cooling of runner until gate has sealed, and then fast cooling of both part and runner! This is not reality; thus, we choose runners which geometrically accomplish low pressure drop and slower cooling; this results in effective fill and pack. We may eject runners which are still soft and molten, but that is OK if we can get runner out of mold. We must take care to size runners as needed - smallest size that will fill and pack part. This is often smaller than designers might think. Mold designers often think the runner design requires a shape for faster cooling such as a true trapezoid with flat bottom and less depth, but the runner's first priority is effective fill and pack of the part. The designs below are useful with softer olefin type resins which might collapse into sprue recess on three plate molds during strip action.

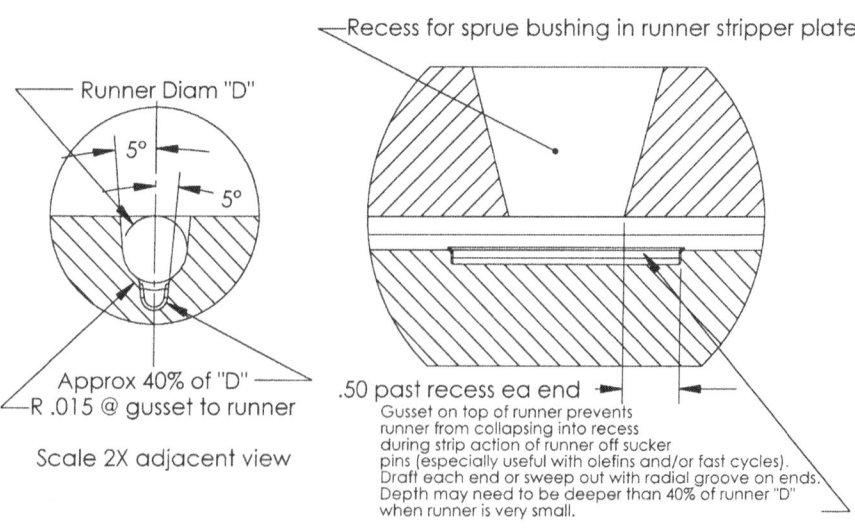

Runner Diam "D"

5°

5°

Recess for sprue bushing in runner stripper plate

Approx 40% of "D"
R .015 @ gusset to runner

Scale 2X adjacent view

.50 past recess ea end

Gusset on top of runner prevents runner from collapsing into recess during strip action of runner off sucker pins (especially useful with olefins and/or fast cycles). Draft each end or sweep out with radial groove on ends. Depth may need to be deeper than 40% of runner "D" when runner is very small.

Runner Sizing

Runner sizing should be done per the following:
A.   Identify maximum part wall thickness.
B.   Size runner by <u>working from part back to sprue</u>.
C.   Set runner adjacent to gate at diameter equal to maximum part wall thickness.
D.   Size each <u>upstream</u> branch by taking the number of branches (typ 2) to the 1/3 power X previous size. NOTE: $2^{1/3} = 1.259$
E.   Repeat previous step until all branches sized.
F.   If primary (largest) runner sizes appear too large and may lengthen the cycle, then oversized runners may be reduced in diameter by a selected percentage. A mold filling analysis can be done to identify fill pressures.
G.   Remember to err on the small or "steel safe" side; it is easier to cut the runner bigger after the first mold trial than to resize smaller.

NUMBER OF BRANCH RUNNERS	RATIO OF FEEDER TO BRANCH DIAMETER
2	1.259
3	1.442
4	1.587
5	1.709
6	1.817
7	1.913
8	2.000

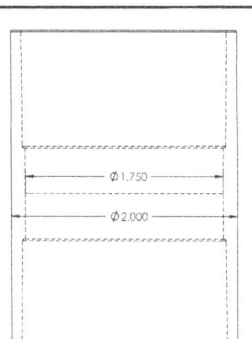

Example: Runner Sizing (Balanced 16 cavity Mold)

Location	Formula	Calculated runner size (inches)
max wall		0.125
A	equal to max wall	0.125
B	=1.259 x 0.125	0.157
C	=1.259 x 0.157	0.198
D	=1.259 x 0.198	0.249

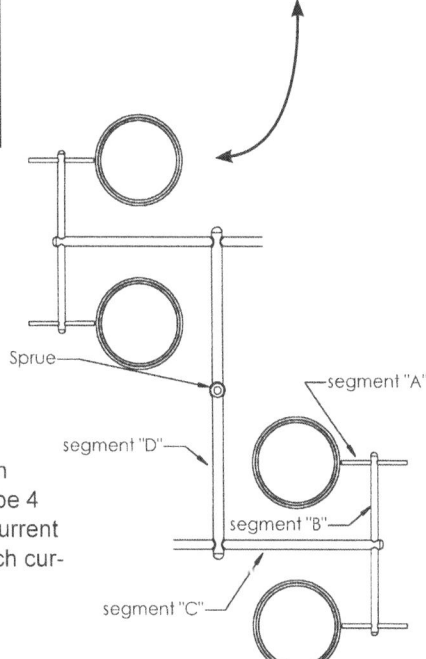

NOTE: As stated above, multiplier is based on # of branches. If the secondary runner branch "C" fed an X pattern supplying four paths (4 branches direct to part); then there would be one less segment and/or size. The segment feeding final X pattern would need to be bigger since there would be 4 branches (0.125 X 1.587=0.198; same as current branch "C" since there would not be a branch currently known as "B").

Weld Lines / Knit Lines

Weld lines can result from the "joint" created when two or more melt flow fronts come together. When the weld lines join, there is often a faint line created. Certain resins and/or colors may make this line more visible. Highly polished surfaces also make these weld lines more visible. Weld lines occur anytime there are multiple gates and/or there are obstructions in the molded part to flow around such as cored holes. Weld lines might also be caused by odd shaped parts lacking symmetry about the gate location or having uneven wall thicknesses. Weld lines are often unavoidable; thus, process conditions and mold design must be such to make weld lines acceptable with regards to both cosmetic and part strength requirements.

The following suggestions are provided to improve weld line appearance or strength:
1. Increase melt temperature. Do not increase to point of degradation or such that increased volatiles are released.
2. Increase injection speed.
3. Increase mold temperature.
4. Improve venting to allow volatiles easier escape from molds. Often there are gaseous volatiles being pushed by the flowing melt front which are trapped at the weld line.
5. Plan gate location so that weld lines may join but must continue to flow; this often results in a somewhat better mix or merge of the two melt fronts.
6. Increase packing or hold pressures.
7. Increase pack or hold time (injection forward time).
8. Decrease the clamping pressure used; this may permit more venting (use caution to avoid flash).
9. A textured mold surface can often serve to hide weld lines which may only be cosmetic problems.
10. Evaluate resin for presence of unneeded additives or lubricants (contact resin supplier).
11. Change gate size and/or location so weld line is moved to a less conspicuous location.
12. Increase part thickness. Note: this may change weld line location.
13. Adjust transfer point and/or transfer method (VP transfer is transfer from velocity control for filling to pressure control for packing; may be done by position, timer, pressure, etc.).
14. Check dryer if molding hygroscopic resin.

Sprue Sticking

Sticking sprues are often solved by any or all of the following (amorphous PET typically requires #7 below):

1. Increase nozzle temperature.
2. Reduce bubbles at base of sprue where it intersects with runner; bubbles can impede sprue pulling by the runner or coldwell; (use increased hold time or lower melt temperature; overpacking may make sprue tighter).
3. Check nozzle orifice ID versus sprue orifice ID: nozzle should be smaller (approx. 1/32" is typical).
4. Check and correct nozzle alignment with locating ring.
5. Remove scratches and grooves caused by steel tools inserted into sprue bushing ID. To correct: Recut with tapered reamer (done by machine shop or tool room).
6. Use backdrafted coldwell or other suitable puller design such as types shown at right.
7. If sprue is large requiring excessive cooling time; the resin must be cooled sufficiently to release from sprue ID. To correct: Make sure sprue bushing is in contact with moldbase steel for full length; eliminate air gaps; improved cooling results from a line on line fit of bushing to hole. Sprue bushings are available in high conductivity copper alloy which helps with faster sprue cooling (better bushing thermal conductivity).

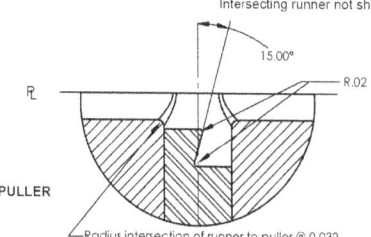

BACKDRAFTED PULLER (PREFERRED)
Intersecting runner not shown in this view
use 3° for gate pullers

Z PULLER
Radius intersection of runner to puller @ 0.032

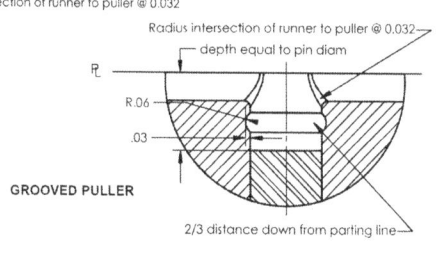

GROOVED PULLER
2/3 distance down from parting line

8. Increase cycle time to accomplish more cooling.
9. Some resins which have poor lubricity and require high packing may benefit from a sprue taper of 3/4 inch per foot (3.57° incl angle) instead of the standard 1/2 inch per foot taper (2.38° incl angle). Do # 7 first.
10. Poor nozzle spherical radius match to sprue bushing's spherical radius. Recut same spherical radius on each (done by machine shop or tool room).
11. Use a nucleating agent which accelerates the setup of crystalline resins. Colored resins typically setup (solidify) faster than do clear or natural resins since the colorant acts somewhat like a nucleating agent.
12. Draw polish sprue ID.
13. Use an electrically heated hot sprue bushing.
14. Try water cooled sprue bushing from HASCO®.

Drool

Drooling problems can be located at the nozzle or at a hot gate in a hot runner system. If the drool is from a hot gate, then the round bubble shape is often called a plastic "BB", since it is shaped as such. These BBs can cause significant mold damage if they have cooled to become hard AND the mold closes before BB removal. The core can force the BB (if larger than wall thickness) into the cavity's gate orifice possibly damaging the thin steel associated with most hot tip gates. See figure below which shows an enlarged view of the gate probe tip and gate well behind cavity.

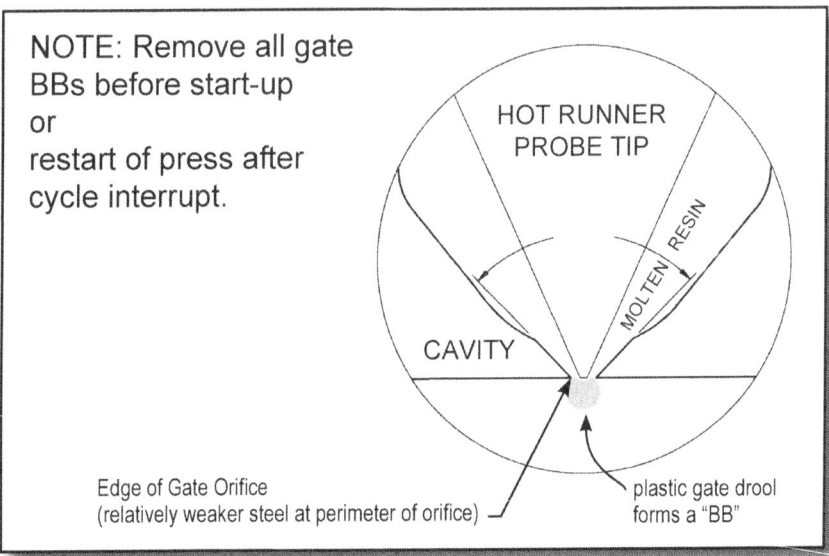

NOTE: Remove all gate BBs before start-up
or
restart of press after cycle interrupt.

HOT RUNNER PROBE TIP

MOLTEN RESIN

CAVITY

Edge of Gate Orifice
(relatively weaker steel at perimeter of orifice)

plastic gate drool forms a "BB"

Corrective actions include:

1. Lower controller set point temperature (drool location).
2. Lower melt temperature (barrel setpoints).
3. Increase melt decompression (suckback). Be watchful for increased splay.
4. Reduce back pressure.
5. Use reverse taper nozzle.
6. Use a nozzle shut-off valve.
7. Clean nozzle filter if used. The use of a nozzle filter impedes the effectiveness of melt decompression. If a filter is used, a cylinder nozzle may be required whereby decompression is accomplished thru sprue break (on a limited basis, since there is less volume potential in the cylinder nozzle versus screw diameter).
8. Moisture in the resin creates major drooling problems; thus, the need for proper drying.
9. Check location of nozzle heater band; proper location near nozzle orifice and ample coverage of nozzle may permit lower nozzle setpoints.

Poor Screw Recovery

Poor screw recovery has the obvious affect on cycle time as well as possible excessive shear heating from increased screw rotation.
Possible corrective action include:

1. Check heater bands for functionality. All heaters should be functional or else screw recovery will likely be slowed and/or hot spots created which may affect balance of fill, etc.
2. Reduce back pressure, but maintain some back pressure for good consistency.
3. Reduce rear zone set point for amorphous resins such as polycarbonate, acrylic, styrene, ABS, SAN & PVC. May also need to increase center zone. The front zone should be set the same as actual melt temperature.
4. Increase rear zone set point for crystalline resins such as polyethylene, polypropylene, nylon, acetal, PET (PET is usually supplied as highly crystalline pellet, but may be amorphous after molding). Crystalline resins require considerable more heat input to change the crystalline structure to amorphous. These guidelines listed in items 3 & 4 are general, the optimum screw recovery time is reached thru experimentation with all barrel setpoints.
5. Vented barrels typically have slower screw recovery since there is a decompression zone in middle which impedes resin conveyance forward. See machine supplier guidelines for optimum barrel profile to minimize the affect of the vented barrel.
6. Check screw and barrel for wear. The clearance is typically specified to be: 0.001 TO 0.0015 inches per side per inch of barrel diameter (e.g. A 2" barrel ID would have a total diametral clearance of 0.004 to 0.006 inches).
7. Check screw for presence of mixing sections which may be no longer needed for current production.
8. Check screw flights for buildup or stuck resin back in feed area. Remove hopper and resin (following your plant's safety guidelines which should include a faceshield), slowly rotate screw backwards and look for buildup (may require manual suckback). The screw should be clean for at least 25% of its length. The normal feed zone is 50% of screw length. Normal resin conveyance forward is accomplished via resin sticking to the hotter barrel and pushed forward by the screw flights to accomplish mixing and thermal homogeneity.

Screw Recovery vs Temperature

Hotter rear zone temperatures may yield faster screw recovery with some resins!

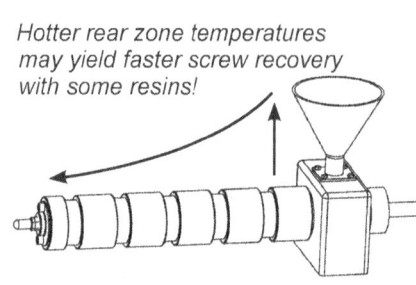

REAR ZONE TEMP (°F)	SCREW RECOVERY (sec)
430	7.95
440	7.49
450	7.07
460	6.66
470	6.41
480	6.32
490	6.28
500	6.31
510	6.40
520	6.51
530	6.67

NOTES:
1. Front zone (and hot runner manifold) should be set same as actual melt temperature.
2. Descending profiles may be beneficial to achieve faster screw recovery.

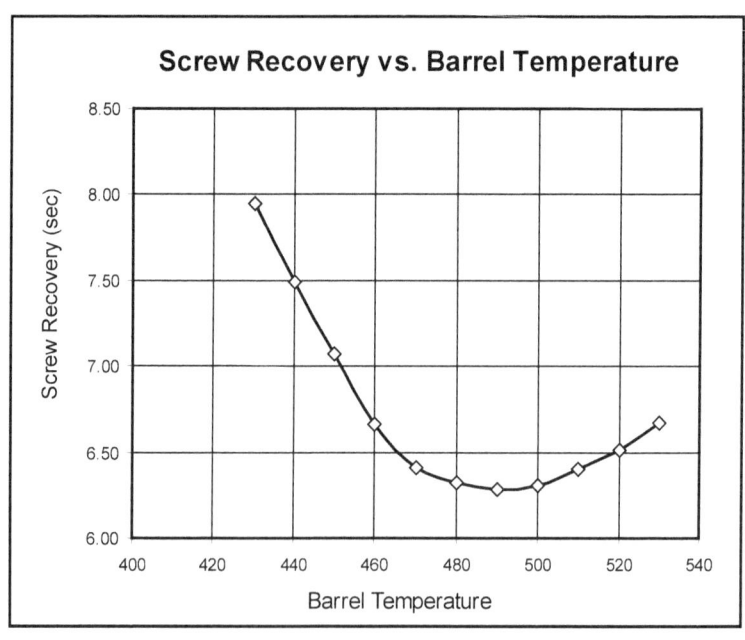

Excessive Cycle Time

Cycle time can be broken into three main components. Cycle reductions can result from a better understanding of these:

1. Fill: faster fill will save small amount of cycle time.
2. Mold open time includes: clamp tonnage release, mold open, eject, mold open idle, mold close and build tonnage. Moving a clamp slowdown switch ½" on mold open/close may yield a ½ - 1 second savings. If mold open time is 20-25% of the total cycle, then purchasing a new state-of-the-art machine will likely be a cost effective improvement. Old machines are often very slow in the following respects:
    a. Slower maximum clamp movement speed.
    b. Less control of the speed such as no dynamic braking whereby clamp movement oil is throttled thru a valve providing true braking action; this permits delayed slowdown switch positions.
    c. Slow to release and build clamp tonnage. Some presses in the 450 ton range require 1.5 to 2 seconds for each of these functions resulting in total mold open times of 7-9 seconds (or more).
    d. Older machines often cannot "eject on the fly" -- ability to eject or pull cores while clamp is moving. Machine hydraulics may not permit multiple functions to occur simultaneously. Ejection on the fly and/or screw rotation during mold opening can improve cycle.
3. Cooling time (includes pack & hold set time + set cooling time). The pack/hold time should accomplish a gate seal then drop off to begin screw recovery. As previously stated, cooling has begun at the start of pack/hold. Cooling can be concluded when (all required):
    a. Screw recovery is accomplished if not doable during mold opening (e.g. older low performance presses).
    b. The part has cooled enough to be reliably ejected from the mold without damage or deformation. A good rule of thumb to use when attempting to develop the fastest cycle is: eject the part at approximately 30° F below the HDT @ 66 psi loading (found in resin properties specifications sheet). Proper draft and polish accelerate when the part can be ejected. See also next page on improved mold cooling.
    c. Part dimensions are as needed. Note: This is affected by original process development; thus, faster cycles should be tested during product/process development when shrink rates are established.

Improving Mold Cooling

Molders do understand that proper voltage and wire size are required to supply equipment; otherwise, we might burn down the plant with an electrical fire. Mold water cooling is also critical, but the consequences aren't dramatic and severe enough to force molders into proper consideration for this cooling. We pay for these mistakes with lost productivity.

The following is suggested toward improved mold cooling:

1. Calculate GPM required (see formula found in the *Injection Molding Reference Guide, 4th Ed* ... also found on pages 82 & 97 of this book). This formula is based on resin, shot wt, temperature and cycle (heat load). Take this calculated requirement times 2.0 or 2.5 for a goal to supply this mold. A value of 1 is the bare minimum required if water flow and heat load were evenly distributed (but they aren't). The 0.5 (if used) allows for future cycle reductions and/or processing at different heat load conditions (melt or eject temperatures). The additional 1 (comprising the 2.0 or 2.5 multiplier) is an arbitrary multiplier in attempt to over supply the circuits so that restrictive circuits with high heat loads (bubblers, baffles, etc. in cores) have enough water. Typical water schematics pass too much water through clamp plates, etc. where there is little or no heat load. Some molders make good use of flow indicator/regulators here to restrict this wasted flow.

2. Do not use quick disconnects with internal shut-off valves -- these are very restrictive and impede flow.

3. Coolant flow requires a pressure differential: keep supply pressure high and return pressures low. This is accomplished by using adequate sizes for hose, fittings and piping.

4. Molds supplied from portable mold heaters should have a properly sized dump valve in this heater/pump unit (¼" is standard, but larger is available).

5. During initial mold qualification, develop a documentation sheet which lists the GPM for each circuit. This must be dynamic which means all circuits must be flowing simultaneously into the normal return line. This is messy and may require a couple hours to quantify, but worth the effort if deficiencies are identified. Once individual flow is known, we can calculate Reynolds number for heat load locations.

6. Review mold designs for cooling improvement opportunities; the cost will often pay for itself later.

Excessive Molded Part Variation

There are many sources of variation in the molding process. Variation can be grouped into one of three main categories (all categories include measurement error; thus, it should be quantified):

A. Within the group (aka within the molded shot). Includes cavity to cavity variation within the shot caused by steel variation or imbalanced fill or cooling differentials in mold.

B. Short term group to group variation (aka shot to shot variation). Includes machine pressure control variation, transfer and cushion variation, coolant temperature fluctuations and inherent resin variation.

C. Long term group to group variation (this also is shot to shot variation). Includes variation listed above plus regrind blend variation, oil temperature changes, barrel temperature controller variation as heaters cycle on and off, and ambient temperature affects.

It is worthwhile to develop a spreadsheet whereby critical dimensions are measured for all cavities for 3-5 (or more) consecutive shots. Then standard deviation and Cpks can be computed for rows and columns for purposes of identifying if most variation is Type A or Type B above. Type C variation is normally quantified in longer term PC studies. Steel measurements should also be reviewed in attempt to explain variation found in the analysis of Type A variation described above.

As previously stated, all data analysis described above includes measurement error; thus, if the same part and dimension is measured 5 times with a total spread of 0.0015 inches than no data is accurate to more than this amount of possible error. It may be useful to perform an established gage repeatability study for very critical measurements. This paragraph serves only to advise those attempting demanding Cpk/Ppk values on dimensions with very close tolerances such as ± 0.001 inches or even ± 0.002 inches. The molding process variation, mold builders variation and measurement error can quickly use the entire tolerance or at least make Cpks beyond 1.33 impractical without extra development cost.

Pages 44 & 45 include some sample data spreadsheets exhibiting how to isolate shot to shot variation from within the shot variation and an example of Ppk improvement. See also the *Injection Molding Reference Guide, 4th Ed.* (and pages 46 & 47 this book) for formulas on Cp, Cpk, Pp & Ppk and standard deviation (σ_{n-1}). It should be noted that Cpk is based on an estimated sigma (R-bar ÷ d2) whereas Ppk uses the true sigma (σ_{n-1}).

Cause and Effect Diagram for Injection Molding

Sources of Variation: Injection Molding

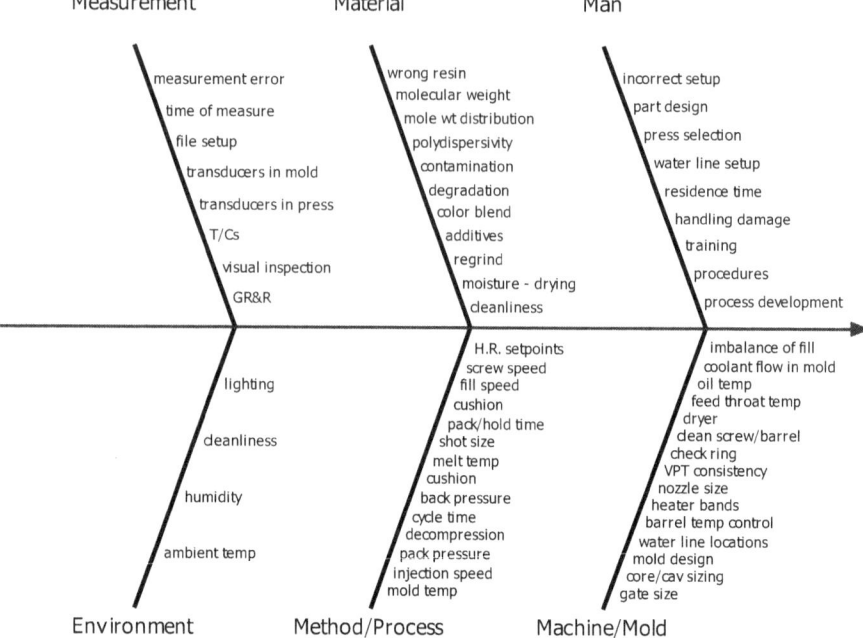

After close examination of the cause and effect diagram above, it can be seen that we as molders only have control over some of the factors effecting variation.

How can we minimize variation from the above:
- procedures & instructions as to mold setup (setup and process used)
- preventive maintenance (PM) to equipment and molds
- gage R&R studies to understand measurement error
- procure good quality equipment and molds
- push SPC upstream to the resin supplier

One significant source of variation for injection molders is the resin variation. We can develop decoupled processes, but the resin variation may ultimately lead to shifts and drifts which greatly challenge our ability to mold products without excessive variation.

As can be seen: there are many process variables in injection molding.

The next page will list various process inputs (setpoints) and process outputs.

Sources of Variation

Check ring – Dirt, metal or other FM (foreign material) can get caught in check ring seat area preventing proper check action which results in resin leakage past check ring – this results in smaller injected shot and reduced packing. Check rings may not check the same every shot.

Resin blend variation – Variation in additive package can cause crystallization or release differences. Molecular weight average or distribution can cause flow viscosity differences and resulting molding variation. Note: The resin supplier has variation just like everyone else.

Resin handling – Change in drying, color blending, regrind blending, contamination, etc causing change to melt viscosity. A drying hopper runs low, but refilled by material handler prior to being empty...this can result in melt viscosity reduction when resin gets to barrel.

Gate blockage – Dirt, metal or other FM may block a small diameter gate; when this happens other cavities may get overpacked because the same plasticized shot gets injected into one (or more) less cavities.

Hyd variation – Flow control and/or pressure valves might be corrupted for a single shot by dirt in valve.

Mold cooling – Water lines can become plugged with debris resulting in poor cooling. Other cooling circuits may have heat transfer impeded by scale or rust. Variation in supply coolant temperature to mold and/or back pressure for return lines may affect actual mold temperature. Variation in mold temperature affects the following: balance of fill, packing (pressure drop) and cooling rate. Incorrect connections such as two ins/zero outs have huge effect on cooling!

Screw wear – Can result in higher melt temperature as plasticizing becomes less efficient resulting in more shear heating.

Electrical spike/noise – May corrupt signal & control for a given shot.

Heater band failure – Machine will likely keep running, but heating and plasticizing characteristics may change depending on which heater and proximity to the thermocouple.

Dirty mold vents – The venting characteristic in a mold may be slowly but constantly changing as the vents get increasingly loaded with volatiles prior to a mold PM. This can affect filling pressure required and volatiles trapped in weld lines.

Ambient conditions – May affect relative humidity which may affect dryer effectiveness; temperature may affect various aspects of machine performance including clamp tonnage, oil temperature, nozzle and barrel temperature. Note also the post mold cooling dynamics affect shrinkage, sink or void developments and/or moisture re-absorption to the resin.

Imbalanced fill – The thermal homogeneity and flow division at branches affects balance of fill and resulting packing, cooling rates and molded-in stress. We can attempt to minimize with consistent screw back pressure and plasticizing RPM. All molds with more than one cavity will have some imbalance of fill.

Closed loop equipment may be partially successful in minimizing the effects of many of these conditions. Closed loop equipment may vary one parameter to achieve consistency with another.

Variation from Cycle Interrupts

The previous page lists many commons sources of molding variation, but it does not list one of the main culprits in causing significant special cause variation -- molding cycle interrupts. These interrupts are <u>not</u> common caused normal variation (even though may be common in some molding plants). The process equilibrium is corrupted in two ways:

1. Longer residence time effects viscosity...usually downward or easier flowing. Lesser viscosity is especially common in the case of thermal sensitive resins like polycarbonate, PET, nylon, acetal, acrylic, etc. This lower viscosity may cause the mold to flash.
2. Mold gets colder because residual heat in the mold is removed. Typically a mold is cooled by a coolant such as water; during normal operation, the molding surfaces have some residual heat that has not been removed completely before next cycle starts. After an interrupt, there is often ample time for the coolant to remove all this residual heat; thus, getting the molding surfaces down to the actual coolant temperature. This colder mold temperature often causes a short shot.

As can be seen, the predicted effects from #s 1 & 2 above are conflicting. It all depends on how thermal sensitive the resin is versus how fast the mold is cycling. If mold cycles fast enough there could be considerable residual heat left behind. The net effect is hard to predict, but the equilibrium is certainly compromised.

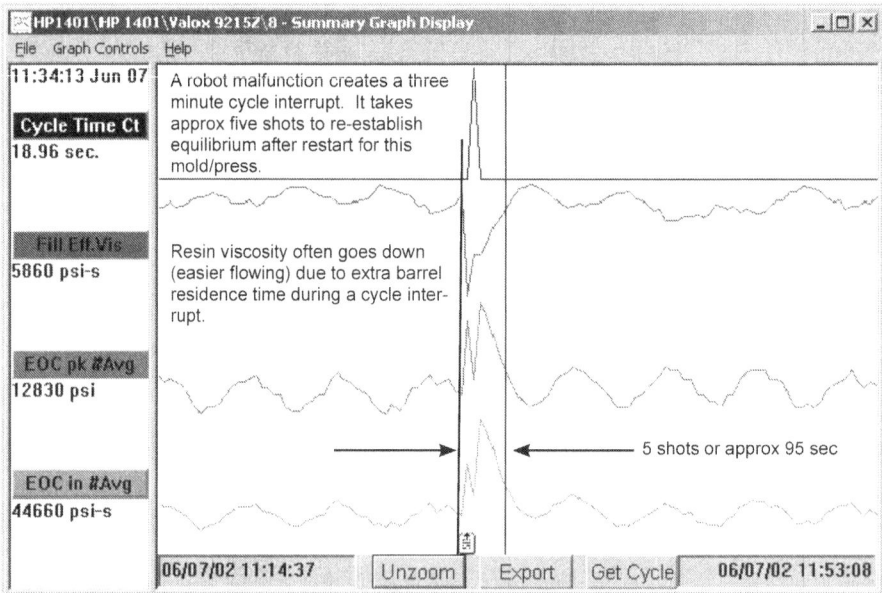

Most good molders today running critical quality type parts have some method to reject the first few start-up shots after an interrupt (these methods are critical to catching significant molded part variation). These methods include the following:

1. Inject a few airshots to remove low viscosity resin -- does not help the colder mold however!
2. Turn off conveyor or robot and manually catch parts.

Root Cause Analysis

Control limit violations should result in some form of root cause analysis and documentation. When performing such root cause analysis, it is worthwhile to understand all the inputs to the process - the many variables affecting the process. Below is a listing of the many items to check, and they are listed in the approximate order of importance and/or ease of checking.

Process Checks for Root Cause Analysis
1. Verify that the part has been remeasured; rule out measurement
2. Proper shot size and resulting cushion
3. Mold temperature control units are on and at proper setpoints
4. Pack/hold pressure is at proper setpoint
5. Barrel temperature setpoints are at proper setpoint
6. Back pressure is turned on and at proper setpoint
7. VP transfer (boost cutoff) is at proper setpoint (and mode thereof)
8. If resin is hygroscopic (meaning - does it require drying):
   a. dryer is at proper setpoint temperature
   b. dryer has not run low in recent hours affecting residence time
   c. dewpoint is properly low (desiccant is good, bed regeneration is good, etc)
   d. dryer and/or material lines are not dirty with contamination
   e. have maintenance check filters, regeneration heaters, desiccant, etc
9. Injection forward timers are at proper setpoint
10. Injection fill velocity is at proper setpoint resulting in proper fill time
11. Cool/cure timer is at proper setpoint resulting in proper cycle time
12. Hot runner setpoints are at proper setpoint
13. Mold is watered correctly
14. Clamp tonnage is sufficient for the mold
15. Resin lot changes having stiffer or easier flow; different thermal properties affecting cooling, warpage, void formation, etc
16. If the above checks good, have maintenance check the press operation: heater bands, hyd. valves, screw & check rings, etc

Tooling Checks for Root Cause Analysis
NOTE: Many of the following can also be caused by process issues.
A. If part sticking or deforming:
   Check for parting line drags and rolled steel creating drags
B. If part is flashing:
   Check for parting line damage, plastic beneath slides, blocked gates
C. If part has FM or metal contamination:
   Check for galling ej pins, other moving steel, vertical shut-offs, etc
D. If gate vestige is poor such as stringing or high gates:
   Check for optimum gate size, check for smashed T/C wires, proper heater size and placement, gate lands, probe tip height for HR tools ... NOTE: depending on gate type, there may be many other possible issues causing poor gate appearances.
E. If part exhibits splay or water spots:
   Check for water leaks caused by damaged o-rings, cracked cores or cavities; check also for hose or pipe nipple leaks whereby the water can migrate thru pipe recesses to steel inserts, then onto molding surface - especially possible with water lines on top of mold.
F. If part exhibits flow lines:
   Check for plastic debris in gate drop from previous shot; check also for sharp corners, engraving, part/tool features affecting plastic flow
G. If part exhibits pin push:
   Check for excessive puller depths, insufficient ejection or poorly placed ejection; check also for air poppets that are not functional.

Shot to Shot vs Within the Shot Variation

Dimension = 1.037 = (Nominal)
Min Tolerance = 1.032
Max Tolerance = 1.042

COLLECT SHOTS EACH 5 MINUTES FOR 2 HOURS; PLACE IN LABELED BAG
RECORD PROCESS AND ANY CHANGES MADE THROUGHOUT THE RUN (THERE SHOULD BE NONE)

CAV	1	2	3	4	5	6	7	8	9	10-22 (not shown)	23	24	AVG	σ_{n-1} STD DEV	Ppk	Pp
1	1.037	1.037	1.036	1.037	1.037	1.036	1.037	1.037	1.036	......	1.037	1.037	1.037	0.00048	3.23	3.46
2	1.038	1.037	1.037	1.037	1.037	1.037	1.038	1.037	1.037	......	1.038	1.037	1.037	0.00041	3.85	4.02
3	1.037	1.037	1.038	1.037	1.037	1.038	1.037	1.037	1.038	......	1.037	1.037	1.037	0.00048	3.23	3.46
4	1.037	1.037	1.038	1.038	1.037	1.038	1.037	1.037	1.038	......	1.037	1.037	1.037	0.00051	2.97	3.27
5	1.039	1.040	1.039	1.038	1.040	1.039	1.039	1.040	1.039	......	1.039	1.040	1.039	0.00066	1.41	2.53
......cavities 6 thru 30 not shown......																
31	1.039	1.040	1.040	1.040	1.040	1.040	1.039	1.040	1.039	......	1.039	1.040	1.040	0.00044	1.70	3.77
32	1.039	1.039	1.039	1.038	1.039	1.038	1.039	1.039	1.039	......	1.039	1.039	1.039	0.00048	2.31	3.46
													AVGs			
AVG	1.038	1.038	1.038	1.038	1.038	1.038	1.038	1.038	1.038	......	1.038	1.038	1.038	0.00060	2.28	2.91
σ_{n-1}	0.0009	0.0012	0.0013	0.0011	0.0012	0.0013	0.0009	0.0012	0.0012	......	0.0010	0.0012	0.00115			
MIN	1.036	1.035	1.035	1.036	1.035	1.035	1.036	1.035	1.035	......	1.036	1.035	1.035			
MAX	1.039	1.040	1.040	1.040	1.040	1.040	1.039	1.040	1.040	......	1.040	1.040	1.040			
n	32	32	32	32	32	32	32	32	32	......	32	32	32			
Ppk	1.55	1.16	1.05	1.21	1.16	1.069	1.55	1.16	1.11	......	1.44	1.14	1.22			
Pp	1.77	1.39	1.31	1.56	1.39	1.33	1.77	1.39	1.37	......	1.66	1.36	1.47			

As can be seen from these averages, the variation is greater within the group or within the shot (or in other words: cavity to cavity variation). The shot to shot variation is less. There is potentially more opportunity to improve Cpk by working on the cavity to cavity variation. Additional study is needed to see if it is steel variation or caused by something else such as imbalanced fill or cooling differentials.

Average of shot to shot variation (between group variation; indicates process consistency) = 0.00060 σ_{n-1}

Average of cav to cav variation (within subgroup variation; may indicate cav steel variation, imbalance fill or temp differentials, etc) = 0.00115 σ_{n-1}

Average Ppk of 24 shots; target is 1.33 or greater = 1.22 Ppk

Calculating and Correcting Ppk

0.400 NOMINAL DIMENSION
0.403 MAX SPEC LIMIT
0.397 MIN SPEC LIMIT

CAV	SHOT 1	SHOT 2	SHOT3	AVG	MIN	MAX	RANGE
1	0.401	0.400	0.400	0.400	0.400	0.401	0.001
2	0.402	0.401	0.402	0.402	0.401	0.402	0.001
3	0.400	0.400	0.401	0.400	0.400	0.401	0.001
4	0.400	0.401	0.400	0.400	0.400	0.401	0.001
5	0.402	0.401	0.401	0.401	0.401	0.402	0.001
6	0.401	0.401	0.401	0.401	0.401	0.401	0.000
7	0.400	0.399	0.400	0.400	0.399	0.400	0.001
8	0.400	0.401	0.401	0.401	0.400	0.401	0.001
9	0.401	0.401	0.400	0.401	0.400	0.401	0.001
10	0.401	0.400	0.401	0.401	0.400	0.401	0.001
11	0.400	0.400	0.400	0.400	0.400	0.400	0.000
12	0.400	0.400	0.400	0.400	0.400	0.400	0.000
13	0.400	0.401	0.400	0.400	0.400	0.401	0.001
14	0.401	0.401	0.401	0.401	0.401	0.401	0.000
15	0.400	0.401	0.401	0.401	0.400	0.401	0.001
16	0.402	0.402	0.401	0.402	0.401	0.402	0.001
17	0.402	0.401	0.401	0.401	0.401	0.402	0.001
18	0.400	0.401	0.401	0.401	0.400	0.401	0.001
19	0.400	0.400	0.401	0.400	0.400	0.401	0.001
20	0.401	0.402	0.402	0.402	0.401	0.402	0.001
21	0.401	0.401	0.402	0.401	0.401	0.402	0.001
22	0.400	0.401	0.401	0.401	0.400	0.401	0.001
23	0.401	0.400	0.401	0.401	0.400	0.401	0.001
24	0.400	0.400	0.400	0.400	0.400	0.400	0.000
25	0.401	0.401	0.402	0.401	0.401	0.402	0.001
26	0.400	0.400	0.401	0.400	0.400	0.401	0.001
27	0.402	0.402	0.401	0.402	0.401	0.402	0.001
28	0.401	0.401	0.402	0.401	0.401	0.402	0.001
29	0.400	0.400	0.401	0.400	0.400	0.401	0.001
30	0.400	0.400	0.400	0.400	0.400	0.400	0.000
31	0.401	0.400	0.400	0.400	0.400	0.401	0.001
32	0.400	0.401	0.400	0.400	0.400	0.401	0.001
AVG	0.4007	0.4007	0.4008	0.4007			0.001
COUNT	32	32	32	32			32
MIN	0.400	0.399	0.400	0.399667			0.000
MAX	0.402	0.402	0.402	0.402			0.001
RANGE	0.002	0.003	0.002	0.002333			0.001
σn-1	0.000745	0.000701	0.000693	0.000713			0.000397
Pp	1.34	1.43	1.44	1.40			
		1.61	1.76	1.55	1.64 after adjustment - Note #4 below		
Ppk	1.05	1.12	1.05	1.072			
		1.32	1.45	1.20	1.32 after adjustment - Note #4 below		

NOTES:

1. Having uniform core & cavity sizing is beneficial from a mold maintenance standpoint; thus, review actual sizing before a change, then verify need for change by review of process and factors contributing to cavity to cavity variation.

2. Comparing three shot cavity averages with nominal spec will indicate location for core or cavity in need of resizing.

3. A Pp significantly greater than the Ppk also indicates potential to improve Ppk.

4. The Ppk (listed as Cpk in earlier versions of this book, but should be called Ppk since using true std dev) above could be improved to average nearly 1.33 by adjusting steel for cavities (2, 16, 20 & 27) downward only 0.001 (ea data point down by 0.001). These were only cavities to average 0.402 in size. Should measure steel to understand root cause!

Cp, Cpk & Pp, Ppk

Cp & Pp are terms used to describe the process potential. The terms Cpk & Ppk are used to describe process capability. The differences between Cp vs Pp or Cpk vs Ppk are in what is used for the standard deviation. There are several terms used to describe standard deviation as follows:

$$s = \text{Sample standard deviation} = \sigma_{n-1} = \sqrt{\frac{\Sigma(X - \bar{X})^2}{n-1}}$$

$$\sigma = \text{Population standard deviation} = \sigma = \sqrt{\frac{\Sigma(X - \bar{X})^2}{n}}$$

$$\hat{\sigma} = \text{estimate of process standard deviation} = \frac{\bar{R}}{d_2}$$

As can be seen s is similar to σ, except for the divisor of n-1 in place of n. Often times, people will use either s or σ_{n-1} to denote the same thing.

The following formulas can be used to compare your actual process performance to the specification limits. These formulas assume the process is in control and data is normally distributed about a central mean or average.

$$\text{Cp} = \frac{\text{USL-LSL}}{6\hat{\sigma}} \qquad \text{Pp} = \frac{\text{USL-LSL}}{6s}$$

When the Cp < 1 the process variation exceeds spec limits. When the Cp > 1 the process variation is less than spec limits, but in either case the Cp (or Pp) only looks at process potential, meaning the actual mean must be centered within the spec limits to make full use of the limits with the process variation that is actually occurring.

It is stated above that Cp is process potential. Cpk (and Ppk) is an indicator of process capability in that it looks at the variation relative to where the mean of the process is located.

$$\text{Cpk (lower)} = \frac{\bar{\bar{X}}\text{-LSL}}{3\hat{\sigma}} \qquad \text{Cpk (upper)} = \frac{\text{USL-}\bar{\bar{X}}}{3\hat{\sigma}}$$

The calculation for Ppk would substitute "s" for the estimated standard deviation -- sigma hat as follows:

$$\text{Ppk (lower)} = \frac{\bar{\bar{X}}\text{-LSL}}{3s} \qquad \text{Ppk (upper)} = \frac{\text{USL-}\bar{\bar{X}}}{3s}$$

Cpk = minimum of either {Cpk(lower),Cpk(upper)}.

Cpk values greater than 1 are favorable. Many companies now require mold/part qualifications to accomplish a Cpk (or Ppk) value of 1.33 to as much as 2.00 depending on the market served. Upon close examination of the Cpk formula, it can be seen that higher values will be obtained by broader spec limits and/or reduced variation -- one makes numerator larger; the other makes denominator smaller, both scenarios result in larger Cpk values.

Cp, Cpk & Pp, Ppk (continued)

The calculation for Cpk (use of sigma hat) is often used in control charts for continuous production. Smaller off-line samples often use the Ppk calculation for ease of calculation. Shown below is a spreadsheet containing aforementioned formulas calculating Ppk and Cpk using s (σ_{n-1}) and σ hat as required.

$$\hat{\sigma} = \frac{\overline{\overline{R}}}{d_2} = \frac{0.00753}{1.128} = 0.00667 \quad \text{see calc next page for s } (\sigma_{n-1})$$

$$Ppk = \frac{\overline{\overline{X}}-LSL}{3s} = \frac{0.7555 - 0.725}{3 \times 0.00720} = \boxed{1.41}$$

$$Cpk = \frac{\overline{\overline{X}}-LSL}{3\hat{\sigma}} = \frac{0.7555 - 0.725}{3 \times 0.00667} = \boxed{1.52}$$

Note: X bar and X double bar are same for this sample and is equal to 0.755 as shown below

1	2	3	4	5	6	7
			(all data in columns 4 thru 7 is from data in column 2, and is used to represent R values in X bar & R charts)			
Nominal	Samples		(n=1 or 2)	n=3	n=4	n=5
0.800 ± 0.075	0.757					
min = 0.725	0.759		0.002			
max = 0.875	0.752		0.007	0.007		
	0.763		0.011		0.011	
	0.758		0.005			0.011
	0.757		0.001	0.006		
	0.755		0.002			
	0.752		0.003		0.006	
	0.756		0.004	0.004		
	0.760		0.004			0.008
	0.735		0.025			
	0.759		0.024	0.025	0.025	
	0.761		0.002			
	0.763		0.002			
	0.751		0.012	0.012		0.028
	0.760		0.009		0.012	
	0.763		0.003			
	0.758		0.005	0.005		
	0.742		0.016			
	0.748		0.006	0.016	0.021	0.021
X bar =	0.7555	R bar =	0.00753	0.01071	0.0150	0.0170
s or σ_{n-1} =	0.00720	σ hat =	0.00667	0.00633	0.00729	0.00731
Pp =	3.47	Cp =	3.75	3.95	3.43	3.43
Ppk =	1.41	Cpk =	1.52	1.60	1.39	1.39
subgroup size =	20		2	3	4	5
		d_2 =	1.128	1.693	2.059	2.326

Note: d_2 is copied from statistical reference tables (included in IM Reference Guide). If we had calculated the Ppk using only first five samples, our Ppk would have been larger at 2.759; when we stop at 10 samples, the Ppk becomes 3.115; when we stop at 15 samples, the Ppk becomes 1.497 -- a large drop down. This large drop down is caused by the value of 0.735 in the data set. This may be measurement error or from some special cause affecting the process. A recheck of this data point would be in order. The point to the previous looks at Ppk with different sample sizes shows that sample size does influence the resulting calculation. This is caused by increased chances for variation with larger sample sizes caused by an increase in number of inspections and more chances for random or special causes of process variation. Larger sample sizes can sometimes improve Ppk (n is in the divisor).

Injection Molding Process Basics

There are many process inputs to the injection molding process. There are some that are highly effective at driving dimensional change; thus, considered as critical process factors.

Before we can set or optimize the critical process factors we must make sure the basics have been satisfied. <u>Almost any process input can be forced to be significant if poorly set</u>; thus, certain basic requirements must set within reasonable guidelines for the process (if any of the following are poorly set, then variation can be greatly increased):

1. Machine must have sufficient clamp tonnage to hold mold closed.
2. Machine must be able to plasticize with time constraints, and accomplish plastication with good melt homogeneity: back psi not too high and not too low (too low results in very inconsistent shot); must have sufficient shot size to yield a small cushion after injection and pack/hold.
3. Machine must have good check ring to transfer shot with pressure into the mold.
4. Mold must have good venting.
5. Machine must be able to inject with adequate fill speed and pressure (especially important for thin wall applications) ... the fill speed greatly effects packing: both <u>fill speed and packing are critical process parameters</u> that will be discussed more later.
6. Machine must be able to eject parts: sufficient ejection stroke and pressure; proper ejection speed to prevent part deformation.
7. Melt temperature must be within supplier guidelines for the resin; <u>melt temperature is often a critical process factor</u>.
8. Mold must be designed with adequate cooling, and circulator pumps must achieve good temperature control and flow rates ... <u>part cooling rate is a critical process parameter</u> (sometimes the part geometry results in uneven cooling rates which requires we respond by varying the mold half temperatures to achieve similar cooling rates between mold halves).
9. Hygroscopic resins must be properly dried.
10. Residence time in dryer and barrel must be within resin supplier guidelines.
11. Nozzle size should be as large as possible, but no larger than sprue orifice in mold.
12. Decompression must be set correct: before or after screw run, too much can cause splay, and too little can result in drool which the next shot must shoot thru causing pressure drop.
13. Feedthroat cooling can effect plasticizing.
14. Clamp open speed can effect part release from A half which could effect part deformation.
15. Resin should be blended properly with regards to additives and colorants.

This list could be expanded with more items such as post mold environment, etc. Item 15 (resin) is a source of variability in injection molding. We frequently forget that the resin supplier has a manufacturing process too that includes variability. Resin variation can be caused by molecular weight variation, weight distribution (broad vs narrow), additives such as nucleating agents, lubes, releases, colorants, thermal stabilizers, etc. <u>There are molding techniques that can help minimize the effects of resin variation</u>.

Basic Process Requirements

- PROPER PLASTICIZED SHOT
  Proper resin .... with good thermal homogeneity .... and proper amount – not too much and not too little! .... not too hot and not too cold! .... no contamination .... proper drying – if required .... etc!

- PROPER PRESS SIZE (especially Shot Size) During injection forward, the machine controls the linear movement of the screw at a set velocity ... if we inject the same volume of plastic over a longer length, we have better resolution of control (e.g. 7 inches vs an oversized press whereby stroke might only be 1 inch ... the ability to transfer using decoupled molding is better with longer stroke).

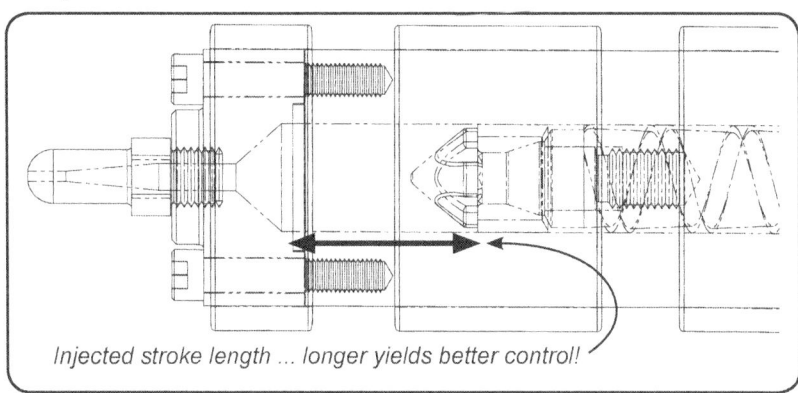

Injected stroke length ... longer yields better control!

- BALANCED FILL
  Needed to avoid momentary stalling of melt fronts – these hesitations cause premature resin cooling and molded in stress and possible cosmetic problems .... not to mention that such stalling and restarting is highly variable resulting in inconsistencies!

- GOOD CHECK RING ON SCREW
  Required for good plasticizing AND shot size control. When we mold with decoupled molding, we have a transfer position which controls the amount of plasticized shot that is NOT injected into the mold. If the check ring does not always check, then we can get too little resin injected into mold which requires too much to be filled on second stage at a likely reduced velocity and pressure which may result in defects .... this often results in the Technician making a shot size adjustment, BUT this can later result in extra material injected into mold when check ring checks sooner vs later which may overpack the mold. We cannot live with sooner or later on check ring performance – there must be consistent and reliable checking of the check ring.

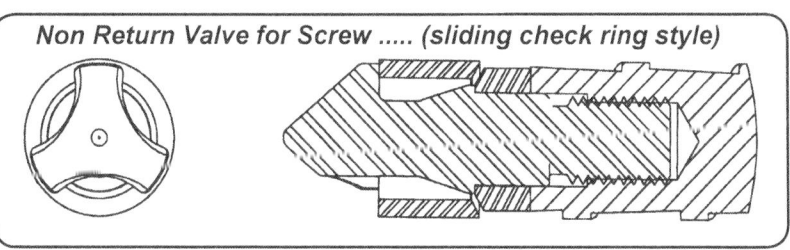

Non Return Valve for Screw ..... (sliding check ring style)

Basic Process Requirements (continued)

- GATE SEAL ACCOMPLISHED
  The gate is like a thermal valve which after cooling becomes closed. We want to inject as long as possible to pack the proper amount of resin into mold. If we stop too soon whereby we stop applying pressure before the valve is sealed, then resin can back flow into the runner; not to mention we have not fully packed the part. If we pack long after the gate is thermally sealed and solidified, we MAY have wasted some time, but cooling does still take place and this time may be needed.....we will have wasted energy of applying pressure. It is best to quantify the gate seal time so you have just a little more time than needed. Even when we have more than enough time, there may be cosmetic problems if the gate is sized too small which results in premature closing of this thermal valve. Mechanical valve gates do not require a gate seal time test as the gate is sealed when you mechanically close the valve, BUT you still must experiment when that time is optimum for best part appearance and dimensional performance. Too long makes it hard to close gate and get good gate appearance....too short of time may not pack enough plastic molecules into mold.

- HIGH QUALITY MOLD
  For proper venting, proper surface finish, proper gate size, good support system to withstand tonnage and injection pressure, good shut-offs to avoid flash, adequate ejection, good cooling, etc, etc, etc!!!!

Four Critical Process Variables

FOUR CRITICAL PROCESS VARIABLES

1. PLASTIC PRESSURE (packing)

Directly effects the number of molecules packed into the cavity, and resulting part size. Adjust to change part size, but must stay inside of "molding window" to avoid flash and short shots. Must also avoid overpacking to the point of causing parts to stick in mold or cause excessive ejection forces whereby part deformation results.

2. PLASTIC FLOW RATE (fill time)

- Faster fill rates result in reduced viscosity.
- Reduced viscosity can result in <u>more consistent</u> viscosity, fill and packing.
- Faster fill rate results in reduced ΔP.
- The reduced ΔP results in more pressure to the cavity resulting in less shrinkage.
- Packing occurs at higher melt temperature which also results in reduced ΔP.

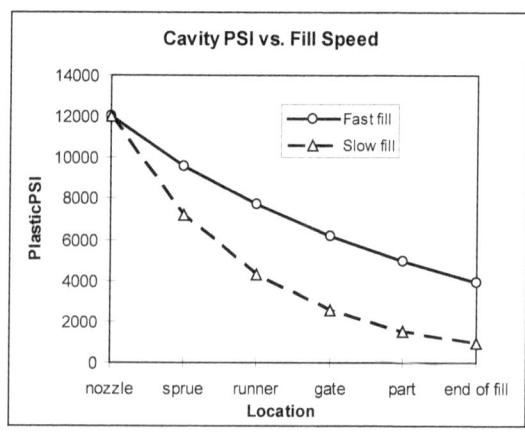

Four Critical Process Variables (continued)

Note: Inject as fast as possible, but reduce as needed to avoid flashing or to correct diesel burns. Should operate molding press whereby fill rate is controlled by velocity setpoints and not pressure limited. This requires fill pressure be set higher than needed, but likely high enough to cause flash if not reduced prior to completing mold fill.

3. PLASTIC COOLING RATE
- Amount of stress retained; slow cooling results in less stress, less retained orientation and higher shrinkage.
- Avoid planned cooling differentials unless needed to overcome warpage. Planned differentials can alter patterns of shrinkage and warpage.
- Economics
- Degree of crystallinity
- If a crystalline resin:
  Slow cooling yields:
  increased crystallinity
  higher tensile strength
  higher stiffness
  less toughness
  Fast cooling yields:
  increased toughness
  faster cycle
  more retained orientation
- Water mold to minimize ΔT, but maximizing flow rates.
- Mold temp is affected by cycle time and interruptions.

NOTE: Cooling time can be calculated.

4. PLASTIC TEMPERATURE
- Hotter yields reduced viscosity and reduced ΔP (viscosity is more affected by fill rate once melted)
- Hotter creates more density change from melt to solid which would result in more shrinkage, but the reduced ΔP will permit improved packing which reduces shrinkage. If added shrink is desired, then increased melt temperature plus increased mold temperatures along with a reduction in pressures would increase shrink.
- Influences amount of retained stress
- Influences amount of crystallinity
- Can affect part cosmetics if too hot or too cold. Best melt temperature is a balance between minimizing heat load to mold resulting in faster cycles and being hot enough for reduced molded in stress. Use supplier mid-range specification as a starting point.
- Influences cooling time (see heat load formula below)
  Heat load = shot wt X specific heat of resin X ΔT
  where
  ΔT = melt temperature minus ejected shot temperature
  This heat load would be for a single molding cycle (1 shot).

Mold Evaluation Options

Debug - This test is a basic check of mold functionality including the following:
- Set and run mold
- Establish a process and record for future reference (includes gate seal)
- Obtain sample parts; distribute as needed
- Perform basic checks (including wall thickness variability for similar walls)
- Suggest tooling improvements needed to improve processability (in writing)

Gate Seal[1] - Should be performed when first establishing the process during debug. Failure to perform this test may cause molding defects associated with under packing the parts such as sinks, excessive shrinkage and warpage. This test will identify the needed injection forward time; when this is estimated it may be estimated too low and cause the aforementioned problems and increase variability if gate does not become sealed.

Balance (Balance of Fill Analysis)[1] - A check of a multi-cavity mold's maximum imbalance of fill (part weight of least filled part versus first filled part; performed as a short shot). Processability is improved and variability is reduced with a more balanced fill.

DOE (Design of Experiments) - A DOE serves to check the mold's (and molded product's) sensitivity to process changes. A DOE done with a one cavity unit mold eliminates the "within group" variation coming from cavity to cavity variation; thus, improving the significance of process variation. With a one cavity mold the important process changes are not clouded by the aforementioned cavity to cavity variation. We want to identify what effect the process has on critical dimensions. The report then serves as a road map to guide processors toward what process variable to change to achieve necessary dimensional compliance.

Fill Time and Pressure Analysis (Relative Viscosity Testing)[1] - This test may be performed when there is concern for the availability of sufficient pressure to fill and pack part on the final production molding press. High pressure and large volumes of oil are needed when filling large cavitation molds. A debug press running a one cavity mold will easily display a performance edge versus a high cavitation mold. A spreadsheet can be arranged comparing the debug presses for unit and production molds versus the end use production press. On the spreadsheet the max fill rate and pressures for a given shot size can be compared. If a very fast fill rate or high filling pressure is required AND the spreadsheet indicates this is NOT doable in the production press, THEN the relative viscosity curve generated in this testing process will indicate where on the viscosity curve the production mold will operate. This information may be needed to plan for a different more suitable press selection OR identify the need to redesign the melt delivery system, gate location or plan a lesser number of cavities in the mold.

2-24 Hour PC Studies - This test would typically be done with the production mold in its final stage prior to normal production. If a required Cpk is planned, then it will be identified during this PC study run. Measurement time can become excessive; thus, a few critical dimensions should be identified for measurement. A common request would be an 8 hour run. This should be done using the planned production molding press.

---

[1] This test is outlined in the *Injection Molding Reference Guide, 4th Ed*; Gate Seal & Balance of Fill can also be seen in this book ... *see pp 66 & 67.*

V-P Transfer (VPT)

The V-P transfer is the transfer from 1st stage filling (under normal machine velocity and pressure controls) to the 2nd stage which is pressure control. Many modern injection molding machines have closed loop control of the injection molding fill rate or fill speed. A linear position sensor is used to track the screw position. The information from the position sensor is fed to a controller which also reads the fill speed setpoints entered by the operator. The controller may ask for higher injection (line) pressures so as to maintain the set/requested injection molding fill speed. The pressure may increase as the resistance to flow increases during mold fill. At a certain point in the injection molding stroke (i.e. when the mold is full or nearly full) the resistance to flow becomes very high and it becomes unrealistic to expect the screw to maintain the desired rate. At this point, control is shifted from being velocity controlled to being pressure controlled: This point is know as Velocity Pressure Transfer or VPT.

Note: If and when the machine becomes pressure limited, there will be no more velocity/speed increases. If the machine's 1st stage pressure is set to 50% and the velocity is set to 85%, we may or may not achieve 85% velocity - it depends on pressure required. As the mold fills and the resistance becomes higher, more pressure is required. The resistance increases as the flow front is constantly being cooled by the mold once it exits the heated barrel and nozzle.

If a car is being driven with the cruise control set at 65 MPH; when the car comes upon a hill, it will begin to automatically depress the accelerator in attempt to maintain the set speed. If the car has low horsepower, eventually the car will be power limited and the set speed will not be maintained. Imagine, how effective the cruise would be if we limited the throttle (and resulting horsepower) to only 35 percent.

A machine should fill under velocity control and not be pressure limited for best consistency. If the machine is not pressure limited, then the velocity control can vary oil volume to achieve the desired fill rate (if closed loop). This does not mean the hydraulic fill pressure must be set to 99 or 100%. We can determine when a reduction in fill pressure begins to increase fill time for a given fill velocity set point. By reducing fill pressure in increments, we can quickly determine how much pressure is required. We would then set the pressure approximately 3-5% higher than needed for the desired fill rate setting and fill time received. This combination may still result in flash if the mold packs at this pressure. We must accomplish a V-P transfer prior to this point of mold flashing. This can be done via the methods described below.

VPT Options:
The control to begin packing pressure (and later hold pressure if needed) requires a signal or switch to accomplish. Changeover at the VPT may be set, or triggered, in the following ways:

1. Screw position - also known as position control (most common today).
2. Hydraulic pressure - also known as pressure control.
3. Cavity pressure control - also known as CPC or as cavity pressure changeover control (works well with today's improved transducers which are more reliable than the transducers when concept was first established).
4. Nozzle pressure - also known as melt pressure (uncommon).
5. Mold opening position (developmental; uses sensor to detect slight separation in mold parting line or clamp opening which would indicate pressure rise in cavity).
6. Timer - common on older machines without control features to do 1 - 5 above. Not possible to separate fill from pack with this method; see also Decoupled Molding℠ discussed on the following pages.

Of these options: 1, 2 and 6 are the most widely used.

NOTE: The term Decoupled Molding℠ includes a service mark by RJG Inc.

Decoupled Molding^{SM}

What is Decoupled Molding^{SM}?

Decoupled molding is simply a setup whereby the molder separates fill from pack: the first injection stage is used only for filling and not for packing. This is checked by setting subsequent stages of pressure to zero and weighing the parts. The shot should appear to be short and weight should be approximately 95% of what the fully packed shot weighs. To accomplish decoupled molding, we must have the machine controls to accomplish the transfer from first stage speed control to second stage pressure: this is usually accomplished by a linear position transducer on the linear stroke of the injection ram (that which drives the screw forward). Our goal is for first stage to be speed control (as fast as mold will permit, but this might be relatively slow) and second stage to be pressure control. This is the basis for most machine control; thus, the term V-P transfer: point of velocity control to pressure control.

Why utilize Decoupled Molding^{SM} techniques?

One advantage of decoupled molding is that the fill may become more consistent by not being pressure limited. We want to have more pressure available than is needed, so that the fill is truly controlled by the velocity set points selected. This velocity often requires more pressure (or a different pressure) than that chosen for packing purposes. When the process is decoupled, the fill is separated from the pack and each can be adjusted without effecting the other (within reasonable limits....large increases in fill speed might cause inertial effects on packing). The other advantage of using decoupled molding is its potential to achieve faster injection fill rates. Decoupled molding allows the molder to set first stage injection pressures higher than would be set otherwise. If "traditional" molding techniques were used whereby the first stage performed fill and pack, and second stage accomplished the hold pressure....the first stage pressure would have to be limited to an amount that does not flash the mold or overpack parts to the point of part damage during ejection. If we are able to use higher hydraulic pressures during fill, then we typically are able to obtain faster injection rates and probably more consistent fill rates. The maximum machine rated fill rate is dependent on pressure being available, which normally is in excess of the desired packing pressure.

There are two main advantages of faster fill rate:

1. Any fill time savings will reduce the overall cycle by that amount.
2. The more important benefit is its effect on resin viscosity during fill. The viscosity of the resin can become significantly less at the high shear rates realized during fast injection rates. As the viscosity becomes less, it also becomes more stable (more stable meaning less variation as seen on a relative viscosity curve).

Why do fast fill rates (high shear rates) reduce viscosity?

The plastic molecules are extremely long relative to their width. Plastic molecules are sometimes described as being like a bucket of worms or spaghetti. This comparison is given primarily to describe the amount of entanglement and randomness of a molten polymer. The entanglement causes the molten polymer to be very viscous. When the polymer is forced to flow quickly by large amounts of pressure, the molecules are forced to align themselves in a more parallel fashion which allows them to flow more easily. This alignment results in a significant drop in the viscosity of the resin (for most resins).

(Service Mark^{SM} of RJG Associates, Inc.)

NOTE: The term Decoupled Molding^{SM} includes a service mark by RJG Associates. The technique described by Decoupled Molding^{SM} is not unique or proprietary to RJG Associates. RJG Associates does promote the use of and offers excellent training in the benefits derived from Decoupled Molding^{SM}.

Decoupled Molding^{SM} (continued)

Potential disadvantages of Decoupled molding

If fill speeds are pushed too fast and/or through too small of orifice, then melt fracture may occur whereby molecular chain length is reduced. The strength of the plastic can be adversely affected if too many chains are broken. Since the transfer switch controls the amount <u>not</u> injected on first stage, the plasticized shot must be consistent and closely controlled by the plasticizing stop switch and the actual plasticizing. A good check ring, screw and barrel are thus required to plasticize the shot and not allow leakage back up the screw flights during injection.

If the mold's shot size is small relative to the machine's injection capacity, then the resolution of control can be very unfavorable. For example: if a 6 inch stroke injects 6 ounces and we move the transfer switch 1/16 inch we have made an adjustment equal to 1/96 of the shot, but if the stroke is 2 inches due to an oversized barrel / shot capacity, then the adjustment equals 1/32 of the shot.

If cavitation is low and a cavity is lost due to gate blockage, then too much resin may exist at the elevated fill pressure for the remaining cavities and an overpack situation will likely result. Consideration must be given to the consequences of overpacking the mold; some molds are adversely affected and others may be more forgiving of overpacking.

PLANNING FOR DECOUPLED MOLDING

The following will make decoupled molding more reliable:
1.   Improve gate reliability (reduce gate self blockages).
     a.   Reduced mold open time (improves hot runner gate's ability to reopen).
     b.   Improve balance of fill.
     c.   Reduce or eliminate gate blockage by contamination.
     d.   Reduce gate drooling or sub-gate vestiges which may cause blockage.
     e.   Reduce thermal differential across mold face (improves balance of fill).
     f.   All gate orifices should be same, as should probe tip locations.
2.   Proper machine sizing and selection.
     a.   Select molding machine so that shot size results in mold using approximately 40-70% of capacity; this will improve linear resolution of control. Note: Avoid having barrel size control cycle due to slow screw recovery.
     b.   Closed loop machines may exhibit better control of the fill rate. This is especially true in setups with large resin variation or presence of regrind. The better fill rate control is also seen after cycle interruptions: the melt temperature and mold temperature are changed significantly during a cycle interruption. If maximum speed is used, verify that the machine/ mold are not pressure limited. <u>If pressure limited, adjust velocity setpoint so that pressure required is less than pressure available</u>.
     c.   A machine with faster injection rates has more performance potential.
     d.   Digitized ram control setpoints are more repeatable and easily set.
     e.   A transfer switch is required to control transfer by screw position, cavity pressure or hydraulic pressure during injection.
3.   Proper machine setup.
     a.   <u>Make plasticizing consistent</u> by using some back pressure and using maximum time available for plasticizing (w/o reducing cycle time).
     b.   Once a good process is established, <u>always run with the same fill time</u> regardless of cavitation. If a setup is good, then each cavity needs to see the same fill time each shot. The viscosity of the resin entering the cavity will be directly related to the fill time.

Resin Process Temperatures

PROCESS MATERIAL	MELT TEMP (°F)	MOLD TEMP (°F)
ABS....med, high impact	440-510	90-180
ABS....high heat grade	510-540	140-180
ACETAL	380-420	140-220
ACRYLIC	420-485	120-180
CA, CAB	360-440	80-130
ETHYLENE VINYL ACETATE	350-400	50-100
IONOMER	420-460	50-100
NYLON....6/6	500-560	120-180
NYLON....6	470-530	100-180
NYLON....6/10	460-510	100-180
NYLON....11	420-480	80-140
NYLON....12	400-450	80-140
POLYCARB....lo/med viscosity	540-570	160-200
POLYCARB....high viscosity	590-640	180-240
POLYESTER - PBT	460-490	100-140
POLYESTER - PET (Bottle)	520-560	60-120
POLYESTER - PETG	480-520	70-120
POLYESTER - PCTG	520-560	70-120
POLYESTER - PCT (GF)	565-590	200-250
POLYESTER - PCTA (GF)	560-590	300-320
POLYETHERIMIDE	680-720	220-300
POLYETHYLENE....low density	340-440	50-100
POLYETHYLENE....med density	390-490	50-120
POLYETHYLENE....high density	420-540	50-150
PPO/STYRENE COPOLYMER	480-580	150-220
POLYPHENYLENE SULFIDE	590-670	190-230
POLYPROPYLENE	420-520	60-150
POLYSTYRENE....G.P.	380-460	50-120
POLYSTYRENE...impact modified	400-480	50-120
POLYSULFONE	650-750	200-320
POLYURETHANE	390-440	70-120
PVC....Flex - Rigid	320-420	50-120
SAN	400-500	120-180
T/P ELASTOMER	340-440	70-120

Notes:
1. Flame retardant grades can be 50° less.
2. Follow manufacturer's guidelines if available.
3. Glass or mineral fillers may require higher heats.
4. Amorphous polyesters may stick to hot steel > 150° F.
5. Long residence time may require lower barrel setpoints.

Calculating Color Blend Ratios

Color concentrates are developed by suppliers to a specific color as requested by the molder. The blend ratio is determined by the concentrate supplier. The following formulas can be used to calculate concentrate required:

Examples below use a 25:1 (Resin -"R" : color - "C") ratio: (e.g. .....for example)

A: Starting with given amount of color concentrate:
Use resin equal to "R" times the weight of concentrate: e.g. 40 lbs of concentrate is available; weigh 25 X 40 or 1000 lbs of resin; combine with the 40 lbs concentrate and blend.

B: Starting with a given amount of resin:
Add color concentrate equal to resin wt divided by "R": e.g. 300 lbs of resin is available; divide 300 by 25 to get 12; add 12 lbs of color to the 300 lbs resin and blend.

C: Blending to achieve a specified total blend weight:
Divide end total weight by ("R"+"C"); this will be the concentrate amount; add resin equal to "R" times the concentrate: e.g. A container holding 300 lbs is to be filled to capacity; 300 divided by 26 determines the amount of color at 11.54 lbs; combine with 11.54 X 25 or 288.5 lbs of resin (The exact total is slightly off 300 due to round off).

D. Blending at the press with screw auger feeders:
Weigh total shot weight and divide by 26; each molding shot or cycle should allow this amount of color to fall. It usually is necessary to shut off the resin hopper to isolate the color from resin to gather a sample of color. Ideally the color is metered out simultaneous to the screw rotation.

NOTES REGARDING PERCENTAGES:

1. When blending to a total weight as in example "C" above, 25:1 is not 96% resin and 4% color, but rather 25/26 or 96.15% resin and 1/26 or 3.85% color.

2. In example "B" above where we start with only the resin or the 300 lbs, 25:1 resin/color could be re-stated as 1:25 color/resin which equals 1/25 or 0.04 or 4% times the resin as the color added to the resin (e.g. 300 X 0.04 = 12 lbs color added, but the 12 lbs color is 3.85% of the total of 312 lbs, 12/312 = 0.0385 or 3.85% of the total blend).

3. If the color is specified as 4% instead of a ratio, we must assume we are blending to the total blend weight whereby we would multiply the total times 0.04, (e.g. 300 lbs total X 0.04 = 12 lbs which leaves 288 lbs of resin; 12+288=300). The ratio here is 96/100:4/100 which reduces to 24:1 (not 25:1 as in earlier examples).

Color concentrate is usually considerably higher in cost than the resin; thus, use only the necessary amount. NOTE: Don't worry about differences above between 25:1 vs 24:1 or 3.85% vs 4.00 % above ... listed to best explain most accurate math ... start with what supplier provides as a blend ratio, then use best fit example above.

Resin Drying Temperatures

MATERIAL	TEMP (°F)	TIME (hrs)	BULK DENS. (lb/ft³)
ABS	190°	3-4	42
ACETAL (HOMOPOLYMER)	200°	1-2	40
ACETAL (COPOLYMER)	210°	1-2	40
ACRYLIC	180°	2-3	42
CELLULOSE BUTYRATE	170°	2-3	39
CELLULOSE ACETATE	170°	2-3	38
CELLULOSE PROPIONATE	170°	2-3	40
IONOMER (SURLYN)	160°	7-8	44
LCP (XYDAR)	300°	3-4	50
NYLON	180°	4-5	41
POLYCARBONATE	250°	3-4	40
P'CARB/PBT/ELAST (XENOY)	260°	3-4	42
PEEK	310°	3-4	52
PET-BOTTLE (EASTAPAK 9921)*	340°	5-6	52
POLYESTER (RYNITE)	275°	3-4	54
PET (THERMX EG001)	265°	4	42
P'ESTER (PBT/PET VALOX 815)	360°	4-5	48
POLYESTER (PBT-VALOX 420)	260°	2-3	48
POLYESTER (PET-VALOX 700)	320°	4-5	48
PETG (EASTAR GNXXX)	155°	6	46
PCTG (EASTAR DNXXX)	165°	6	45
PCTG/P'CARB (EASTALLOY DAXXX)	200°	4-6	43
PCT (THERMX CG907)	160°	4-6	41
PCT (THERMX CG921)	160°	6	50
PCTA (THERMX AG230)	330°	4	44
POLYARYLATE	250°	5-6	50
POLYETHERIMIDE	310°	4-5	52
POLYETHYLENE (BLACK)	160°	3-4	34
PPO/STYRENE (NORYL)	210°	2-3	49
POLYPHENYLENE SULFIDE	280°	2-3	50
POLYSULFONE	275°	3-4	50
POLYURETHANE	180°	2-3	48
SAN	180°	3-4	40
SMA (DYLARK)	200°	2-3	38
TPE (HYTREL)	210°	2-3	48
TPR (SANTOPRENE)	160°	2-3	46

* Can dry crystallized PET bottle resin at lower temps for longer time (i.e. 8 hrs @ 295° F).

NOTES:
1. Process return air should be cooled to 150° F or below. The desiccant will be more efficient with increased affinity for moisture at temps below 150°.
2. Fines have a higher ratio of surface area to mass; thus, will absorb moisture faster and more readily than pellets.
3. Some dryers require as much as 4 hours to regenerate; thus, to be considered when dryer is first turned on. Cool down of the bed regeneration should be with previously dried process air so as not to load the desiccant with moisture.
4. Calculate and allow for proper residence time.
5. Clean filters on a regular basis.
6. Excessive drying may result in color shift (some resins, especially clear).

Calculating Dryer Residence Time

Many resins require drying prior to molding to remove either surface moisture or absorbed moisture from the pellets. In order to calculate residence time in the dryer we must calculate the dryer weight capacity. A given dryer has an internal volume, but will hold different weights of resin depending on that resin's bulk density.

First we must gather the knowns so as to calculate the unknowns:

Knowns:
Dryer hopper is 5.7 cu feet of capacity
We can weigh a sample or consult the resin supplier to determine the resin's bulk density; in this example it will be 40 lbs/ft^3.
The molding cycle is 28.5 seconds and the shot size is 188.4 gr.

We must first determine the molding press thruput rate or total pounds per hour processed.

$$\frac{3600 \frac{sec}{hr}}{28.5 \frac{sec}{shot}} \times \frac{188.4 \frac{gr}{shot}}{1} \times \frac{1 lb}{453.6 \, gr} = Y$$

$$\frac{126.32}{hr} \times \frac{188.4}{1} \times \frac{1 lb}{453.6} = Y$$

$$52.47 \frac{lbs}{hr} = Y$$

We convert 5.7 ft^3 dryer capacity to pounds:

$$5.7 \, ft^3 \times 40 \frac{lbs}{ft^3} = 228 \, lbs$$

We can now divide 228 lbs by 52.47 lbs/hr to get dryer residence time in hours:

$$\frac{228 \, lbs}{52.47 \frac{lbs}{hr}} = Y$$

$$4.35 \, hrs = Y$$

It is useful to reexamine the above calculation to review how we cancel units. Sometimes it is easy to forget the units, but this can result in incorrect conversions and incorrect answers if we do not perform all necessary steps to make units come out right. If the units are wrong we will have a numerical answer that is wrong for the intended calculation.

Setting and Starting The Mold

MOLD SETUP:

1. Determine resin requirements & availability; clean hopper, magnet, loader; load resin; start dryer if required.
2. Locate proper KO bars - all having equal length.
3. Prepare platens & mold using stone and mineral spirits.
4. Select eyebolt hole which yields a level hang/lift.
5. Do NOT stand below hanging mold; avoid hitting tie bars when lowering mold into place.
6. Line up locating rings; slowly close mold.
7. Level mold if not already and clamp to fixed platen.
8. Open moving platen (w/ hoist still attached/supporting mold); install KO bars. If KOs are acting as pullbacks: tighten bars making certain they bottom out against the ejector plate in mold.
9. Close platen; clamp mold to moving platen; remove safety straps; unhook hoist.
10. Open mold to desired daylight; set slowdown switches with certainty that banging the mold will not occur; fine tune the final switch positions by repetitive small adjustments, observations and readjustment.
11. Secure KOs to ejector plate of press; set stroke.
12. Connect all required power - hyd, electric, pneumatic.
13. Make sure powered functions are functional; run electrical heaters just long enough to prove functionality avoiding excessive heat buildup before water is connected.
14. Connect water lines using an acceptable number of loops/jumpers; locate lines clear of any interference. Avoid having all "INS" on the same side.
15. Recheck fittings for proper connection; turn water on; (heaters should be off); look for leaks.

PROCESS SETUP (IF UNKNOWN):

16. Set barrel profile per resin supplier's recommended mid-range (same logic for mold temp).
17. Estimate the shot size and set machine for approximately 2/3 of the mold's full shot. Set decompression stroke. Set a position transfer point (if machine is so equipped) approximately one inch from bottom.
18. Estimate & set second stage time; set second stage pressure at zero.
19. Set 1st stage pressure at 50% for starters (this may ultimately be set at 100% - assuming Decoupled Molding^{SM}).
20. Set velocity to maximum.
21. Estimate and set cooling time.
22. Set back pressure at 50 psi.
23. Refer also to Hot Runner Start-up procedure if applicable.

MOLD START-UP:

24. Purge barrel free of degraded resin.
25. Set machine for semi-auto; start cycle; observe screw.
26. Adjust velocity and/or pressure as needed; if the fill was fast and short as estimated, the pressure can be increased. The fill pressure should be set high enough so the fill speed is not pressure limited, but controlled by velocity setpoints. If flash or dieseling occur, slow the velocity.
27. After observing each cycle, the shot size and transfer point will be adjusted frequently to set the process so that the first stage accomplishes 95 - 98 % of the fill as measured by shot weight.
28. Once the first stage shot size, transfer, velocity and pressure are set, we can set 2nd stage packing pressure.
29. Adjust pack pressure as needed, but do not overpack.
30. Recheck cushion; some cushion should be maintained.
31. Set screw rpm so recovery is completed just prior to next cycle, but not limiting cycle time.

PROCESS DOCUMENTATION:

32. Record all basic machines setpoints on the setup sheet.
33. Note the transfer time (fill time) and weight.
34. Note the overall cycle time.
35. Note the ejection: multiple, push only, push/pull, etc.
36. Total shot weight, part weight, % runner, etc.

IMPORTANT NOTE: When using Decoupled Molding^{SM} (see also pages discussing this technique): Considerable skill and specific mold and machine knowledge is required when setting pressures near maximum. Set pressures in accordance with consideration for mold damage in the event some parts do not shoot due to gate blockage and remaining cavities actually see the elevated 1st stage pressure.

Hot Runner Startup & Operation

Hot runner systems offer the following benefits to the molder: elimination or reduction of regrind; automation of part handling is easier; potential for faster cycles; part/runner separation is avoided; injection pressures may be lower. It is important to follow certain guidelines to get the most out of your hot runner system.

Some mold makers will list min and max temperature differentials for the manifold system versus the mold temperature; these should be followed as they allow for proper preload from thermal expansion: This is important to prevent leakage and/or excessive compression damage. A heat soak time of approximately thirty minutes should be allowed after set points are reached to permit proper thermal expansion (assumes resin w/ good thermal stability ... some will require less heat soak time). Front clamping plate cooling is necessary to prevent excessive heating of the platens. It is worthwhile to start-up the system with all zones off; then check each zone by itself to determine that the T/C response is for the respective zone heating. Always make a cavity sheet which indicates which zone number is what cavity and what section of mold relates to which manifold zone. Set the manifold temperature equal to the process melt temperature. Adjust drops as needed to achieve proper gate vestige. Chilled molds need to be started at room temperature. Do not connect gate coolant water in series. If your controller doesn't have soft start or ground fault capabilities; set the temps at only one hundred degrees for the first half hour so as to dry the absorbed moisture out of heater insulation. If a filter nozzle is used, great care is needed in filter nozzle selection to avoid excessive pressure drops - both on injection and ability to decompress manifold before mold opens. Always perform a balance of fill analysis before the mold leaves the mold maker; insist on a maximum imbalance of ten to twenty percent (depending on cavitation). It is best to never purge through the open hot runner mold due to potential for leakage and potential damage to cavities (i.e. some systems which have multiple cavities in a single cavity block could develop forces greater than retention screws retaining cavities or plate strength). It is necessary to sometimes purge through an open mold to purge degraded resin: try to purge with back pressure or minimal injection pressures; never use operating pressures to purge through the open mold. Never connect/disconnect plugs which are carrying power as it can damage most controllers besides being unsafe.

Torque Specifications For Fasteners

GENERAL TORQUE SPECIFICATIONS ENGLISH FASTENERS (FOOT-POUNDS)						
MATERIAL GRADE / BOLT SIZE	SAE 2 MILD STEEL	SAE 5	SAE 8	SHCS	BRASS	SS AISI 303
1/4-20	6	11	12	13	5	5
1/4-28	7	13	15	16	6	7
5/16-18	13	21	25	27	8	9
5/16-24	14	23	30	33	9	10
3/8-16	23	38	50	52	15	17
3/8-24	26	40	60	60	16	18
7/16-14	37	55	85	86	23	25
7/16-20	41	60	95	95	25	28
1/2-13	57	85	125	130	32	37
1/2-20	64	95	140	145	34	40
9/16-12	80	125	175	180	44	50
9/16-18	91	140	195	210	48	54
5/8-11	111	175	245	255	68	75
5/8-18	128	210	270	290	73	80

GENERAL TORQUE SPECIFICATIONS METRIC FASTENERS (NEWTON METERS)							
MATERIAL CLASS / MM-DIAM	4.6	4.8	5.8	8.8	9.8	10.9	12.9
5	3	4	5	7	8	11	12
6	5	6	8	12.5	14	17	20
6.3	5.5	8	9.5	14	16	21	24
8	12	16	20	30	34	44	50
10	23	32	40	60	70	85	100
12	40	56	70	103	120	150	180
14	65	90	110	167	190	240	280
16	100	140	170	270	290	380	440
18	137	177	225	350	--	480	580
20	200	--	330	520	--	740	860

Note: check also torque recommendations from your fastener supplier and/or equipment/product manufacturer for item which fasteners are being used. These values are approximate. SHCS used for Husky, Moldmaster and other hot runner systems may use slightly lower torque values; see supplier guidelines.

SHCS Dimensions (TYPICAL 1960 SERIES)

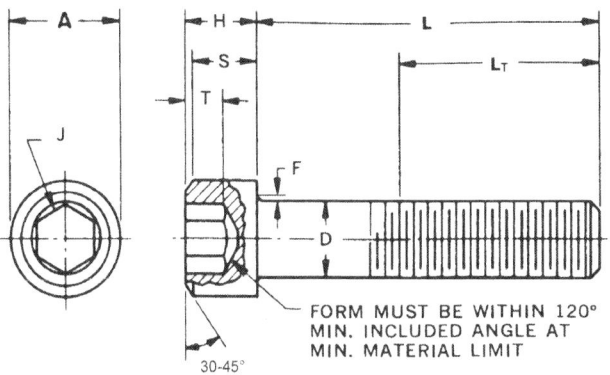

FORM MUST BE WITHIN 120°
MIN. INCLUDED ANGLE AT
MIN. MATERIAL LIMIT

30-45°

SIZE	D (MAX)	A (MAX)	H (MAX)	S (MIN)	J (NOM)	T (MIN)	F (MAX)	L (MIN)
0	0.060	0.096	0.060	0.054	0.050	0.025	0.007	0.50
1	0.073	0.118	0.073	0.066	1/16	0.031	0.007	0.62
2	0.086	0.140	0.086	0.077	5/64	0.038	0.008	0.62
3	0.099	0.161	0.099	0.089	5/64	0.044	0.008	0.62
4	0.112	0.183	0.112	0.101	3/32	0.051	0.009	0.75
5	0.125	0.205	0.125	0.112	3/32	0.057	0.010	0.75
6	0.138	0.226	0.138	0.124	7/64	0.064	0.010	0.75
8	0.164	0.270	0.164	0.148	9/64	0.077	0.012	0.88
10	0.190	0.312	0.190	0.171	5/32	0.090	0.014	0.88
1/4	0.250	0.375	0.250	0.225	3/16	0.120	0.014	1.00
5/16	0.312	0.469	0.312	0.281	1/4	0.151	0.017	1.12
3/8	0.375	0.562	0.375	0.337	5/16	0.182	0.020	1.25
7/16	0.437	0.656	0.438	0.394	3/8	0.213	0.023	1.38
1/2	0.500	0.750	0.500	0.450	3/8	0.245	0.026	1.50
5/8	0.625	0.938	0.625	0.562	1/2	0.307	0.032	1.75
3/4	0.750	1.125	0.750	0.675	5/8	0.370	0.039	2.00
7/8	0.875	1.312	0.875	0.787	3/4	0.432	0.044	2.25
1	1.000	1.500	1.000	0.900	3/4	0.495	0.050	2.50

Note:
1. 1960 series SHCS are typically made from a high grade alloy steel, hardened to a range of 37-45 RC.
2. "F" above is a fillet extension beyond "D".
3. Consult supplier to determine available lengths "L" for each screw size. Typical length increments are as follows:
   1/16" increments - lengths 1/8" thru 1/4"
   1/8" increments - lengths 1/4" thru 1"
   1/4" increments - lengths 1" thru 3.5"
   1/2" increments - lengths 3.5" thru 7"
   1" increments - lengths 7" thru 10"

New Mold Debug Checklist

EASE OF MOLD LIFTING AND SETTING:
1. Will mold fit the end use molding press?
2. Mold should have eyebolt holes & safety strap.
3. Mold should hang level to align with locating ring & KOs.
4. Mold should offer protection to external wiring, switches, fittings, etc on bottom of mold.
5. Mold should have clamping slots positioned such that sufficient clamps can be used to secure mold to platens.
6. All water fittings should be positioned so that not in interference with mold clamps, machine doors, tie bars, etc.
7. Mold should have sufficient eyebolts so that mold rotation can be achieved with a second eyebolt and hoist.
8. Sprue radius should be compatible, sprue recess ID should be compatible with machine's nozzle/heater band OD.
9. Mold should have total weight listed on clamping plate.
10. Socket head cap screws retaining clamping plate should be slightly below flush to prevent platen damage.
11. Electrical receptacles should be accessible & away from water fittings.

BASIC MECHANICAL FUNCTIONALITY:
12. Water lines should be free flowing; check with flow meter/indicator or direct return into bucket to observe flow (do not try to measure/quantify with bucket - not accurate due to lack of back pressure).
13. Ejector plate should return freely and completely.
14. Slides should move freely, but be retained during mold open in the pulled position.
15. Leader pins should not exhibit galling.
16. All moving slide surfaces and slide locks should be free from galling.

ELECTRICAL FUNCTIONALITY:
17. Know the heater wattage; calculate the full load amperage and compare to actual (or at least compare to other like heaters).
18. Each T/C should indicate a proper response for it's heater.
19. Heater cables should be grounded at mold & controller ends.

LOCATION ITEMS:
20. Ejector pins should be flush to 0.001" below flush.
21. Examine parts looking for long ejector pins.
22. Examine parts/runner looking for parting line mismatch.
23. Examine parts looking for slide mislocation.
24. Check part for proper wall thickness with pointed micrometers.
25. Examine wall thickness variation relative to fill problems.
26. Intentional drags should be adjacent to ejector pins and be shallow @ 0.003", but sharp (can then reduce sharpness if too much drag).
27. Alignment should be located on horizontal and vertical centerlines.

ESTABLISH A BASIC PROCESS:
28. Record all pertinent process & setup data.
29. Determine gate seal time.

DETERMINE BALANCE OF FILL:
30. Establish basic process.
31. Set hold/pack pressure to zero.
32. Adjust inject time or transfer position to achieve only one full part (or as close as possible).
33. Collect 3 shots; separate parts by cavity and determine avg part weight for each cavity.
34. Max imbalance should be 15% or less.

MISC OTHER CHECKS:
35. Have all sharp corners been removed from mold base exterior?
36. Do parts fall free and clear of mold without knicks?
37. Have pry bar slots been installed in main parting lines?
38. Review gate vestige relative to requirements.
39. Review for proper mold surface finish from both cosmetic and functionality standpoint.
40. Lightly touch each core and cavity just after repeated cycling looking for hot spots (Do not touch hot molds; observe standard safety practices before reaching into press).
41. Stop mold prior to any ejection, look for raised surfaces where part may be trying to stick in non-moving half.
42. Look at parts for parting line drags.
43. Perform dimensional checks after shrinkage (48 hrs).

Process Development Press Comparisons

Spreadsheet example to compare press used for process development vs press used for production (in this example, it is unit mold press vs. production mold).

	Resin Melt Density	Cavs	Total Runner Weight (grams)	Part Weight (grams)	Shot Weight (grams)	Shot Weight (% of mach)	Machine Shot Size (grams)	Available Injection rate cc/sec	Fastest Fill Time Possible (sec)	plast rate grams/sec	Fastest Plast. Time Possible (@70%) (sec)	max psi avail	Mech Adv psi	Cycle (sec)	Resin Use Lbs/hr
End Use Production Press / Part															
NISSEI 460 TON PET barrel	1.2	16	15.82	23.446	390.956	21%	1871	401	0.81	85	6.57	23127	11.32	16.60	184.88

Cycle breakdown planned PET barrel

Fill	1.8
Pack & Hold	3.3
Cool incl Plasticize	6
Open	5.5
	16.6

Fastest Fill: **OK**

Fastest Plast.: **Problem**

Color Use Lbs/hr @ 1.00% = 1.87

> This type comparison is essential when first establishing a process so that consideration is made to ensure the established process is doable in the planned production press (with respect to fill time, plasticizing time, and packing pressures). Note also: Hot runner molds should have the ΔP thru system quantified by supplier so that the production press has pressure for HR system ΔP and the normal process pressure required (this consideration is important for comparing unit mold to production mold). A unit mold is a one cavity pilot mold for product and tooling development.

	Resin Melt Density	Cavs	Total Runner Weight (grams)	Part Weight (grams)	Shot Weight (grams)	Shot Weight (% of mach)	Machine Shot Size (grams)	Available Injection rate cc/sec	Fastest Fill Time Possible (sec)	plast rate grams/sec	Fastest Plast. Time Possible (@70%) (sec)	max psi avail	Mech Adv psi	Unit press maximum % to not exceed prod press
Sampling Press / Part														
NESSEI 120 TON PET barrel	1.2	1	0.962	23.446	24.408	11%	230	151.8	0.13	16.9	2.06	26170	13.08	88%

CELL G5=(D5*F5)+E5.......CELL H5=G5/I5.......CELL K5=(((F5*D5)+E5)/C5)/J5.......CELL M5=G5/(L5*0.7).......CELL P5=B13......

CELL Q5=((3600/P5)*G5*(1-Q$11))/454 ....CELL K9=IF($B$9>=$K$5, "OK", "PROBLEM").......CELL M11=IF($B$11>=$M$5, "OK", "PROBLEM") ....

CELL Q12=((3600/P5)*G5*Q$11)/454.......CELLS K21 & M21 CAN BE COPIED FROM K5 & M5

Determining Gate Seal Time

Knowledge of the gate seal time, permits the molder to use the needed injection forward time to accomplish effective packing. Injection forward times less than the gate seal time often result in sinks, voids and increased shrinkage resulting in poor dimensional compliance. The total injection forward time should be set a fraction longer than the gate seal time if cycle time permits. Hot tip gates may not result in a seal due to localized packing around the hot gate. This packing is of little or no benefit; thus, do not attempt a full gate seal on some hot gates parts. Note also: valve gates are sealed mechanically... this test applies differently to valve gated molds; whereas, it can still indicate optimum injection time, but valve must be closed in time to achieve good gate vestige.

Make a plot of total part weight vs increasing increments of pack and/or hold time after VPT as shown in table at right.

Increments can be ½ or 1 sec ... or finer increments as needed. Bigger parts with bigger gates need bigger increments of time.

When the part weight becomes stable, the gate is sealed and no more pack/hold time is needed.

Unlike the Balance of Fill test on next page ... here the graph (below) is very helpful to understand the data!

Hold Time (sec)	Shot Weight (parts only) (grams)
0.60	8.716
0.70	8.770
0.80	8.810
0.90	8.830
1.00	8.850
1.10	8.863
1.20	8.878
1.30	8.885
1.50	8.902
1.70	8.913
1.80	8.915
2.00	8.916
2.20	8.916
2.40	8.916

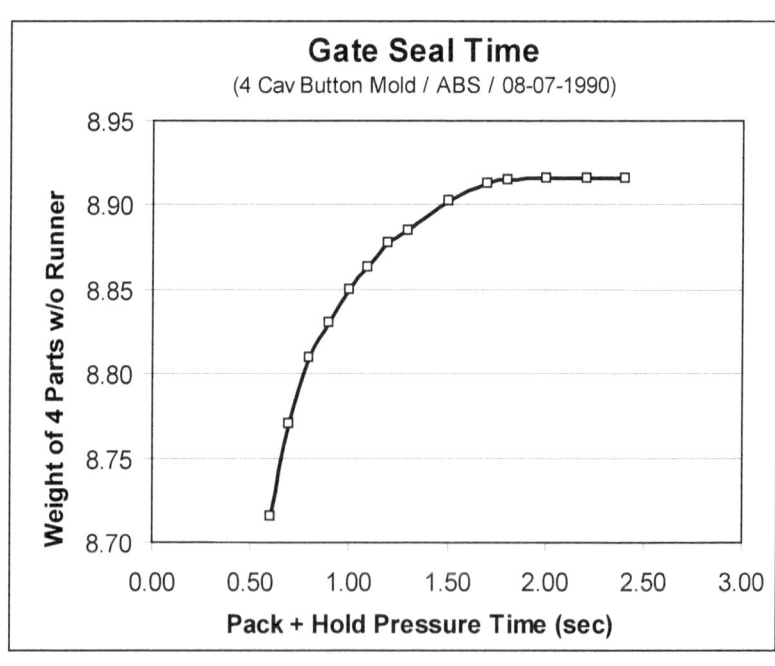

Gate Seal Time
(4 Cav Button Mold / ABS / 08-07-1990)

Balance of Fill Analysis

Cavity	Avg Part Wt (grams)	Fill Seq	Avg % Unbal	% Full	Part wts (grams)		
					shot 1	shot 2	shot 3
4	1.40	1	0.00%	100.00%	1.42	1.39	1.38
5	1.39	2	0.72%	99.28%	1.39	1.39	1.38
12	1.38	3	1.19%	98.81%	1.39	1.37	1.38
13	1.38	4	1.43%	98.57%	1.37	1.39	1.37
6	1.37	5	1.91%	98.09%	1.38	1.37	1.36
14	1.34	6	3.82%	96.18%	1.34	1.35	1.34
11	1.33	7	4.77%	95.23%	1.36	1.32	1.31
3	1.33	8	5.01%	94.99%	1.35	1.32	1.31
15	1.19	9	14.56%	85.44%	1.20	1.21	1.17
10	1.15	10	17.66%	82.34%	1.16	1.15	1.14
9	1.14	11	18.38%	81.62%	1.15	1.15	1.12
7	1.12	12	19.57%	80.43%	1.12	1.14	1.11
1	1.11	13	20.29%	79.71%	1.13	1.11	1.10
8	1.06	14	23.87%	76.13%	1.10	1.05	1.04
16	1.06	15	23.87%	76.13%	1.07	1.06	1.06
2	1.03	16	26.01%	73.99%	1.04	1.03	1.03

PROCEDURE:

1. Process should be running at equilibrium with regards to mold & melt temperature, pressures normal & stable.
2. Consider mold's ability to run a short shot.
3. Set feed or transfer to run short shots whereby one part is full (or nearly so); begin shot collection. Typically, this will be 1st stage fill only as we are only concerned with balance of FILL ... we do not need any pack or hold psi in this test.

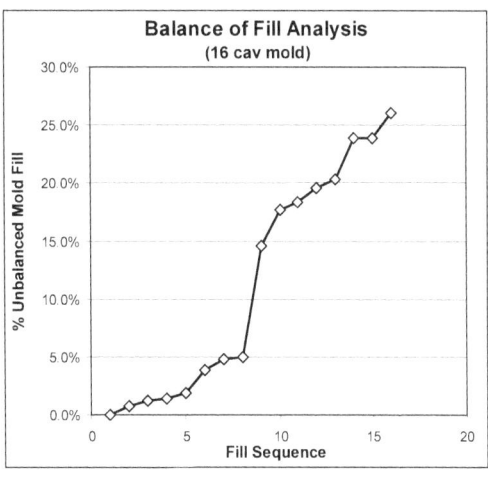

Balance of Fill Analysis (16 cav mold)

4. Collect three shots separated by cavity number, weigh individual parts and record, calculate average.
5. Arrange data in table, sorted in descending order (from high to low) of average part weight.
6. Compute % unbalanced (this mold has a 26% imbalance - see table above).
7. Restore process to normal.
8. Graph serves no purpose other than display of data.

Sample Mold Information Sheet

Mold Information Sheet

MOLD DESCRIPTION					
NUMBER OF CAVS.		MOLD DRWG NUMBER			
PART DESCRIPTION					
RUNNER TYPE					
MOLD SIZE	HORIZ	VERT		SHUT HT	WT

SPRUE RADIUS		SPRUE ORIFICE		
RECESS I.D.		RECESS DEPTH		
EJECTION TYPE				
K.O. PATTERNS				
K.O. THREAD				
GATE SIZE/TYPE				
CAVITY MATERIAL		HARDNESS		Rc
CORE MATERIAL		HARDNESS		Rc
POWER	MANIFOLD			
CONSUMPTION (watts)	GATE DROPS			

SEE CAVITY LOCATIONS BELOW
TOP/EJECTOR HALF   00

OPERATOR SIDE

13	9		5	1
14	10		6	2
Manifold 2			Manifold 1	
15	11		7	3
16	12		8	4

Adopt a standard cavity numbering sequence such as top to bottom starting at offset corner (back side, top in this example).

Since the HR system is located in A-half, the layouts typically start in top left when looking at parting line; thus, this layout. This view is B half where you grab the parts!

Sample Mold Process Sheet

Molding Conditions Record

DATE/TIME: _____ PART: _____ PART WT (GRAMS): _____
PART DESCRIPTION: ___ _____ MOLD NO: _____ SHOT WT (GRAMS): _____
REV # OR ECN (IF APPLICABLE): _____ NO. CAVITIES: _____
MACHINE ID: _____ TONS: _____ SHOT SIZE (OZ): _____
MECH ADV: _____ SCREW TYPE: _____ NOZZLE ORIFICE ID: _____
RUN BY: _____
CORE SEQUENCE: _____ CIRCLE EJECT TYPE: PUSH ONLY PUSH/PULL PUSH/SPRING RETURN OTHER

	RUN NO. OR ACTIVITY DESCRIPTION >>>>>							
MATERIAL	MATERIAL							
	LOT NO.							
	COLOR CONC							
	COLOR BLEND RATIO							
	OTHER ADDITIVES							
DRYING	DEW POINT	(°F) (°C)						
	TEMPERATURE	(°F) (°C)						
	TIME	(HOURS)						
	HOPPER SIZE							
	CALCULATED NORMAL RESIDENCE TIME							
TEMPERATURES	FEED ZONE	(°F) (°C)						
	CENTER ZONE	(°F) (°C)						
	FRONT ZONE	(°F) (°C)						
	NOZZLE	(°F) (°C)						
	HOT MANIFOLD	(°F) (°C)						
	ACTUAL MELT.	(°F) (°C)						
	MOLD - FIXED	(°F) (°C)						
	MOLD - MOVEABLE	(°F) (°C)						
	MOLD - SLIDES	(°F) (°C)						
PRESSURE	FILL VELOCITY	(%)						
	CLAMP	(%) (TONS)						
	FILLING PRESSURE	P1						
	PACKING PRESSURE	P2						
	HOLD PRESSURE	P3						
	BACK PRESSURE							
CYCLE TIMES	INJ. TOTAL SCREW FORWARD	(SEC)						
	FILL TIME	(SEC)						
	PACK TIME	(SEC)						
	HOLD TIME	(SEC)						
	COOLING	(SEC)						
	PLASTICIZING	(SEC)						
	MOLD OPEN TIMER	(SEC)						
	ACTUAL TOTAL OPEN	(SEC)						
	OVERALL	(SEC)						
	RESIDENCE TIME - BARREL (MIN.)	CALC						
MISCELLANEOUS	TRANSFER METHOD							
	TRANSFER POSITION	(INCHES) (MM)						
	TRANSFER WEIGHT	(%)	CALC					
	DECOMPRESSION	(INCHES) (MM)						
	SHOT SIZE	(INCHES) (MM)						
	CUSHION	(INCHES) (MM)						
	END OF STROKE	(INCHES) (MM)						
	SCREW SPEED	(RPM) (%)						
COMMENTS	A -							
	B -							
	C -							
	D -							
	E -							

TESTING PRODUCT SENSITIVITY TO PROCESS VARIATION

It is beneficial to perform designed experiments (also known as DOEs or DOXs.... they mean the same thing). The purpose of a DOE can be to check a mold's (and molded product's) sensitivity to process changes. DOEs are excellent for trouble-shooting dimensional non-compliance, warpage, sink and other molding problems. Identifying such sensitivity yields the following benefits:

- Identify process variables having most affect on dimensional performance; can then make planned process adjustments to maintain or achieve dimensional compliance.
- Can use data as guide in establishing or selecting a single process to run by. This single process would be used in production.
- Identify how "robust" or forgiving the mold is to process changes.

There are several analysis techniques which are associated with Design of Experiments. One of the more common is known as ANOVA which is short for analysis of variance. On subsequent pages you will see a simple table of averages to analyze the effects.

A goal of Engineering is to design the product, mold, process, etc. to be more forgiving or more robust which means less sensitive to the many sources of variation.

FACTORS AND LEVELS

Prior to running the experiments, the process parameters are reviewed and selected based on process knowledge. We often have a feel for which parameters are key to influencing certain responses. In the injection molding industry, there are four process factors which are useful toward evaluating product sensitivity to process variation and resulting product performance.

- melt temperature (± approx 15°- 25° F from nominal)
- injection speed or fill time (± approx 30% from nominal, if possible)
- packing pressure (± approx 100 psi hydraulic from nominal, if possible)
- mold temperature (± approx 10°- 20° F from nominal)

NOTE: IF mold temp differentials are tested; <u>limit max difference to $\cong 30°$ F (a max 20° F ΔT would be better, but 30 may be doable – review mold for near vertical shut-off issues)</u>.

The selection of these parameters assumes reasonable selection of packing time, and the many other process variables which are easily set, but can have adverse effects if poorly set. The machine must also be stable and exhibit an "in control" pattern of operation. We may select only 3 or 4 of these factors, and we may decide on two mold temps to test mold temperature differentials. We should always include packing pressure and fill time.

The process parameters listed are called "factors" when selected for use in the designed experiment. After selecting the factors, the "levels" must be selected. If melt temperature is selected as one factor then the levels are decided; they can be low and high or specific temperatures. If low and high are used, the low and high must eventually be identified and maintained throughout the experiment.

REPLICATION AND ORDER

We must also decide on "replication" and "order". The replication indicates how many times process combinations will be tested (if at all). Replication can be partial or multiple; whereby, the partial is faster, but data is incomplete (not all process combinations are tested). Multiple means running the same combinations of factors and levels more than once to achieve higher confidence in the results. "Order" refers to the order in which the selected conditions (combination of factors) are run; these should be randomized. It is often useful to reorder so that melt temperature is only changed once. It is very time consuming and difficult to know when true equilibrium is reached with regards to melt temperature changes. The thermocouple may indicate that the change has been accomplished, but it is not located in the melt stream and temperature controllers may overshoot the set point before resuming normal control where heaters cycle on and off on a regular basis.

RESPONSES

The dimension or physical property, etc. that is to be measured or quantified as a result of the experiment is termed the "response". Responses are typically variables type data - meaning a numerical value recorded from a measurement of a dimension or test of strength, etc.

Attribute data is simply pass/fail, good/bad or go/no-go types of data. Attribute data such as good or bad appearance with regards to sinks can be coded on a scale of 1-5 for example to be converted into variables type data for inclusion into the experiment data analysis.

PREPARATION

Before running a DOE, it is necessary to discuss the DOE for purposes of selecting the responses, the factors and the levels. This is typically best accomplished by discussing the DOE with those most familiar with the mold, machine, resin and the process. It is also advisable to experiment with selected parameters to verify that all combinations of factors and levels are doable without resulting in short shots, flash or otherwise unacceptable products. Experienced processors can usually identify the combinations which will interact to reflect the most extreme process condition for preliminary tests of feasibility. It should be noted however, that the actual conclusions as to which variables have most effect on responses are the result of statistical analysis on the DOE data. Prior to running the DOE, you should also have already:

- Established the balance of fill as being acceptable.
- Determined the needed overall cycle time.
- Determined the pack & hold pressure limitations.
- Determined the gate seal time (select an overall injection forward time).

DURING THE DOE

- Label all bags with the standard order number (Design ID number).
  Collect 3 - 5 shots for each set of process conditions. Do not close bags until residual heat is released, do not pack bags too tightly.
- Keep all process conditions, resin, machine, etc. as constants; change only the selected factors.
- Record all constant conditions for future reference.
- If making temperature changes, wait sufficient time to reach equilibrium after temperature change is made before gathering new samples (15 minutes after new indicated set point is achieved on mold temperature changes; 45 minutes after new indicated set point is achieved on melt temperature changes). We typically renumber the run order to limit melt temperature changes to one; this is done because of potential variation introduced by premature collection of samples before a melt temperature change fully reaches equilibrium (barrel T/Cs may not fully indicate if and when equilibrium has been reestablished).
- Make notes of observations made during each set of conditions.
- Make effort to complete DOE on same day, start to finish, if possible to minimize variation.

AFTER THE DOE

- Allow parts to stabilize with regards to shrinkage (typically 48 hours).
- Measure responses: typically critical dimensions, gram weight, etc.
- Input data into a suitable software program and analyze per software guidelines. If special DOE software is not available, then spreadsheet software or manual calculations can be used as demonstrated on the following pages.
- Save parts for future reference.
- Issue report with results and conclusions.

Consider the following as a simple experiment. We want to test what effect the two factors - injection fill time and injection pack pressure - have on diameter of the part (the diameter is also known as the response). If we decide on full replication, then all combinations of factors will be run once. The total number of combinations is determined by taking the number of levels to an exponential value equal to the number of factors.

For example:
2 factors = 2^2 = 4 ... (2levels$^{2\ factors}$ = 4 combos) ...
4 factors = 2^4 = 16 ... (2levels$^{4\ factors}$ = 16 combos)

FACTOR		LEVELS	
FILL TIME		LO(-) HI(+)	
PACK PSI		LO(-) HI(+)	

Process Combination	1	2	3	4
fill time	LO(-)	LO(-)	HI(+)	HI(+)
pack psi	LO(-)	HI(+)	LO(-)	HI(+)
response	2.495	2.502	2.478	2.483

Based on this data it is not clear which factor has the most pronounced effect for this response. See example below for a simple method to analyze and quantify the relative strength of the factors for this experiment.

PRIMARY EFFECTS		INTERACTION EFFECTS	RESPONSE
fill time	pressure	time x pressure	
LO(-)	LO(-)	(+)	2.495
LO(-)	HI(+)	(-)	2.502
HI (+)	LO(-)	(-)	2.478
HI(+)	HI(+)	(+)	2.483
2.4805	2.4925	2.489	<<<(avg at HI or +)
2.4985	2.4865	2.490	<<<(avg at LO or -)
-0.018	0.006	-0.001	<<<relative effect (HI minus LO)

The "avg at HI" effect by inj time is 2.4805 which comes from the average of 2.478 & 2.483 (see shaded areas to identify which responses added to compute average).

The aforementioned data is better handled by a spreadsheet formatted as follows (especially true for 3, 4 or more factor experiments; see next page).

Proc Comb	Factors & Levels			Response	Fill	Pack	Interaction
	Fill	Pack	X				
1	-1	-1	1	2.495	-2.495	-2.495	2.495
2	1	-1	-1	2.478	2.478	-2.478	-2.478
3	-1	1	-1	2.502	-2.502	2.502	-2.502
4	1	1	1	2.483	2.483	2.483	2.483
				sum total >	-0.036	0.012	-0.002

compare results ... see also next page

The fill time has 3 times more effect on the response diameter than does pack. NOTE The differences above between results for fill time: -0.018 vs -0.036 ... the top method of calculation is better where we take average at high minus average at low vs a simple sum of the column ... see also next page for further discussion. In example above, the fill is 3x stronger either way of calculation.

You can easily construct a table of averages for your DOE analysis like shown below to analyze relative effects ... in example below the temps had little effect relative to pack psi and fill time ... pack was strongest factor!

	A	B	C	D	E	F		H	I	J	K
	Process Combination	Levels				Response		Fill	Pack	A Temp	B Temp
		Fill	Pack	A Temp	B Temp						
4	1	-1	-1	-1	-1	0.818		-0.818	-0.818	-0.818	-0.818
5	2	1	-1	-1	-1	0.811		0.811	-0.811	-0.811	-0.811
6	3	-1	1	-1	-1	0.832		-0.832	0.832	-0.832	-0.832
7	4	1	1	-1	-1	0.821		0.821	0.821	-0.821	-0.821
8	5	-1	-1	1	-1	0.814		-0.814	-0.814	0.814	-0.814
9	6	1	-1	1	-1	0.807		0.807	-0.807	0.807	-0.807
10	7	-1	1	1	-1	0.832		-0.832	0.832	0.832	-0.832
11	8	1	1	1	-1	0.817		0.817	0.817	0.817	-0.817
12	9	-1	-1	-1	1	0.817		-0.817	-0.817	-0.817	0.817
13	10	1	-1	-1	1	0.812		0.812	-0.812	-0.812	0.812
14	11	-1	1	-1	1	0.832		-0.832	0.832	-0.832	0.832
15	12	1	1	-1	1	0.820		0.820	0.820	-0.820	0.820
16	13	-1	-1	1	1	0.814		-0.814	-0.814	0.814	0.814
17	14	1	-1	1	1	0.808		0.808	-0.808	0.808	0.808
18	15	-1	1	1	1	0.829		-0.829	0.829	0.829	0.829
19	16	1	1	1	1	0.815		0.815	0.815	0.815	0.815
20								-0.077	0.097	-0.027	-0.005
21								-0.010	0.012	-0.003	-0.001

Rows 20 & 21 indicate relative effects, but calculated two different ways: e.g. cell I20 is a simple sum of I4 to I19, but cell I21 is the average of HIs minus avg of LOs ... same relative effect, but clustered closer to zero in row 21 so values near zero are known to be little or no effect. minus values are indirect relationship such as smaller fill time yields large parts; whereas, larger pack yields larger parts.

CELL I21 = SUMIF(I4:I19,">0")/((COUNT(I4:I19)/2))-ABS(SUMIF(I4:I19,"<0")/((COUNT(I4:I19)/2)) written this way so can copy to cells H21 thru K21

BAR to PSI Conversions

psi	bar	psi	bar	psi	bar	psi	bar
15	1	740	51	1465	101	2190	151
29	2	754	52	1479	102	2204	152
44	3	769	53	1494	103	2219	153
58	4	783	54	1508	104	2233	154
73	5	798	55	1523	105	2248	155
87	6	812	56	1537	106	2262	156
102	7	827	57	1552	107	2277	157
116	8	841	58	1566	108	2291	158
131	9	856	59	1581	109	2306	159
145	10	870	60	1595	110	2320	160
160	11	885	61	1610	111	2335	161
174	12	899	62	1624	112	2349	162
189	13	914	63	1639	113	2364	163
203	14	928	64	1653	114	2378	164
218	15	943	65	1668	115	2393	165
232	16	957	66	1682	116	2407	166
247	17	972	67	1697	117	2422	167
261	18	986	68	1711	118	2436	168
276	19	1001	69	1726	119	2451	169
290	20	1015	70	1740	120	2465	170
305	21	1030	71	1755	121	2480	171
319	22	1044	72	1769	122	2494	172
334	23	1059	73	1784	123	2509	173
348	24	1073	74	1798	124	2523	174
363	25	1088	75	1813	125	2538	175
377	26	1102	76	1827	126	2552	176
392	27	1117	77	1842	127	2567	177
406	28	1131	78	1856	128	2581	178
421	29	1146	79	1871	129	2596	179
435	30	1160	80	1885	130	2610	180
450	31	1175	81	1900	131	2625	181
464	32	1189	82	1914	132	2639	182
479	33	1204	83	1929	133	2654	183
493	34	1218	84	1943	134	2668	184
508	35	1233	85	1958	135	2683	185
522	36	1247	86	1972	136	2697	186
537	37	1262	87	1987	137	2712	187
551	38	1276	88	2001	138	2726	188
566	39	1291	89	2016	139	2741	189
580	40	1305	90	2030	140	2755	190
595	41	1320	91	2045	141	2770	191
609	42	1334	92	2059	142	2784	192
624	43	1349	93	2074	143	2799	193
638	44	1363	94	2088	144	2813	194
653	45	1378	95	2103	145	2828	195
667	46	1392	96	2117	146	2842	196
682	47	1407	97	2132	147	2857	197
696	48	1421	98	2146	148	2871	198
711	49	1436	99	2161	149	2886	199
725	50	1450	100	2175	150	2900	200

NOTE: Numbers rounded to nearest whole number;
1 bar = 14.504 psi; 1 kg/cm^2 = 14.223 psi

Temperature Convert °F to °C (001-200 °F)

°F	°C	°F	°C	°F	°C	°F	°C
1	-17	51	11	101	38	151	66
2	-17	52	11	102	39	152	67
3	-16	53	12	103	39	153	67
4	-16	54	12	104	40	154	68
5	-15	55	13	105	41	155	68
6	-14	56	13	106	41	156	69
7	-14	57	14	107	42	157	69
8	-13	58	14	108	42	158	70
9	-13	59	15	109	43	159	71
10	-12	60	16	110	43	160	71
11	-12	61	16	111	44	161	72
12	-11	62	17	112	44	162	72
13	-11	63	17	113	45	163	73
14	-10	64	18	114	46	164	73
15	-9	65	18	115	46	165	74
16	-9	66	19	116	47	166	74
17	-8	67	19	117	47	167	75
18	-8	68	20	118	48	168	76
19	-7	69	21	119	48	169	76
20	-7	70	21	120	49	170	77
21	-6	71	22	121	49	171	77
22	-6	72	22	122	50	172	78
23	-5	73	23	123	51	173	78
24	-4	74	23	124	51	174	79
25	-4	75	24	125	52	175	79
26	-3	76	24	126	52	176	80
27	-3	77	25	127	53	177	81
28	-2	78	26	128	53	178	81
29	-2	79	26	129	54	179	82
30	-1	80	27	130	54	180	82
31	-1	81	27	131	55	181	83
32	0	82	28	132	56	182	83
33	1	83	28	133	56	183	84
34	1	84	29	134	57	184	84
35	2	85	29	135	57	185	85
36	2	86	30	136	58	186	86
37	3	87	31	137	58	187	86
38	3	88	31	138	59	188	87
39	4	89	32	139	59	189	87
40	4	90	32	140	60	190	88
41	5	91	33	141	61	191	88
42	6	92	33	142	61	192	89
43	6	93	34	143	62	193	89
44	7	94	34	144	62	194	90
45	7	95	35	145	63	195	91
46	8	96	36	146	63	196	91
47	8	97	36	147	64	197	92
48	9	98	37	148	64	198	92
49	9	99	37	149	65	199	93
50	10	100	38	150	66	200	93

Temperature Convert °F to °C (201-400 °F)

°F	°C	°F	°C	°F	°C	°F	°C
201	94	251	122	301	149	351	177
202	94	252	122	302	150	352	178
203	95	253	123	303	151	353	178
204	96	254	123	304	151	354	179
205	96	255	124	305	152	355	179
206	97	256	124	306	152	356	180
207	97	257	125	307	153	357	181
208	98	258	126	308	153	358	181
209	98	259	126	309	154	359	182
210	99	260	127	310	154	360	182
211	99	261	127	311	155	361	183
212	100	262	128	312	156	362	183
213	101	263	128	313	156	363	184
214	101	264	129	314	157	364	184
215	102	265	129	315	157	365	185
216	102	266	130	316	158	366	186
217	103	267	131	317	158	367	186
218	103	268	131	318	159	368	187
219	104	269	132	319	159	369	187
220	104	270	132	320	160	370	188
221	105	271	133	321	161	371	188
222	106	272	133	322	161	372	189
223	106	273	134	323	162	373	189
224	107	274	134	324	162	374	190
225	107	275	135	325	163	375	191
226	108	276	136	326	163	376	191
227	108	277	136	327	164	377	192
228	109	278	137	328	164	378	192
229	109	279	137	329	165	379	193
230	110	280	138	330	166	380	193
231	111	281	138	331	166	381	194
232	111	282	139	332	167	382	194
233	112	283	139	333	167	383	195
234	112	284	140	334	168	384	196
235	113	285	141	335	168	385	196
236	113	286	141	336	169	386	197
237	114	287	142	337	169	387	197
238	114	288	142	338	170	388	198
239	115	289	143	339	171	389	198
240	116	290	143	340	171	390	199
241	116	291	144	341	172	391	199
242	117	292	144	342	172	392	200
243	117	293	145	343	173	393	201
244	118	294	146	344	173	394	201
245	118	295	146	345	174	395	202
246	119	296	147	316	174	396	202
247	119	297	147	347	175	397	203
248	120	298	148	348	176	398	203
249	121	299	148	349	176	399	204
250	121	300	149	350	177	400	204

Temperature Convert °F to °C (401-600 °F)

°F	°C	°F	°C	°F	°C	°F	°C
401	205	451	233	501	261	551	288
402	206	452	233	502	261	552	289
403	206	453	234	503	262	553	289
404	207	454	234	504	262	554	290
405	207	455	235	505	263	555	291
406	208	456	236	506	263	556	291
407	208	457	236	507	264	557	292
408	209	458	237	508	264	558	292
409	209	459	237	509	265	559	293
410	210	460	238	510	266	560	293
411	211	461	238	511	266	561	294
412	211	462	239	512	267	562	294
413	212	463	239	513	267	563	295
414	212	464	240	514	268	564	296
415	213	465	241	515	268	565	296
416	213	466	241	516	269	566	297
417	214	467	242	517	269	567	297
418	214	468	242	518	270	568	298
419	215	469	243	519	271	569	298
420	216	470	243	520	271	570	299
421	216	471	244	521	272	571	299
422	217	472	244	522	272	572	300
423	217	473	245	523	273	573	301
424	218	474	246	524	273	574	301
425	218	475	246	525	274	575	302
426	219	476	247	526	274	576	302
427	219	477	247	527	275	577	303
428	220	478	248	528	276	578	303
429	221	479	248	529	276	579	304
430	221	480	249	530	277	580	304
431	222	481	249	531	277	581	305
432	222	482	250	532	278	582	306
433	223	483	251	533	278	583	306
434	223	484	251	534	279	584	307
435	224	485	252	535	279	585	307
436	224	486	252	536	280	586	308
437	225	487	253	537	281	587	308
438	226	488	253	538	281	588	309
439	226	489	254	539	282	589	309
440	227	490	254	540	282	590	310
441	227	491	255	541	283	591	311
442	228	492	256	542	283	592	311
443	228	493	256	543	284	593	312
444	229	494	257	544	284	594	312
445	229	495	257	545	285	595	313
446	230	496	258	546	286	596	313
447	231	497	258	547	286	597	314
448	231	498	259	548	287	598	314
449	232	499	259	549	287	599	315
450	232	500	260	550	288	600	316

Temperature Convert °F to °C (601-800 °F)

°F	°C	°F	°C	°F	°C	°F	°C
601	316	651	344	701	372	751	399
602	317	652	344	702	372	752	400
603	317	653	345	703	373	753	401
604	318	654	346	704	373	754	401
605	318	655	346	705	374	755	402
606	319	656	347	706	374	756	402
607	319	657	347	707	375	757	403
608	320	658	348	708	376	758	403
609	321	659	348	709	376	759	404
610	321	660	349	710	377	760	404
611	322	661	349	711	377	761	405
612	322	662	350	712	378	762	406
613	323	663	351	713	378	763	406
614	323	664	351	714	379	764	407
615	324	665	352	715	379	765	407
616	324	666	352	716	380	766	408
617	325	667	353	717	381	767	408
618	326	668	353	718	381	768	409
619	326	669	354	719	382	769	409
620	327	670	354	720	382	770	410
621	327	671	355	721	383	771	411
622	328	672	356	722	383	772	411
623	328	673	356	723	384	773	412
624	329	674	357	724	384	774	412
625	329	675	357	725	385	775	413
626	330	676	358	726	386	776	413
627	331	677	358	727	386	777	414
628	331	678	359	728	387	778	414
629	332	679	359	729	387	779	415
630	332	680	360	730	388	780	416
631	333	681	361	731	388	781	416
632	333	682	361	732	389	782	417
633	334	683	362	733	389	783	417
634	334	684	362	734	390	784	418
635	335	685	363	735	391	785	418
636	336	686	363	736	391	786	419
637	336	687	364	737	392	787	419
638	337	688	364	738	392	788	420
639	337	689	365	739	393	789	421
640	338	690	366	740	393	790	421
641	338	691	366	741	394	791	422
642	339	692	367	742	394	792	422
643	339	693	367	743	395	793	423
644	340	694	368	744	396	794	423
645	341	695	368	745	396	795	424
646	341	696	369	746	397	796	424
647	342	697	369	747	397	797	425
648	342	698	370	748	398	798	426
649	343	699	371	749	398	799	426
650	343	700	371	750	399	800	427

Tips, Tricks & Shortcuts

1. Check nozzle contact with sprue bushing using 0.002" brass shim stock; leaves good impression of contact area.

2. Check wall thickness in parts with pointed micrometers; saw parts so that the position of pieces does not become lost; mold component misalignment will result in thin walls (radiused corners, tapered walls).

3. Profiling the injection speed can be used to alter the fill pattern on imbalanced molds.

4. Never change gate size to balance fill.

5. Mold clamps should be long enough so that bolts are no more than halfway from mold to fulcrum point of clamp.

6. Leave screw forward on shutdown; otherwise start-up requires barrel to heat slug of plastic in front of screw, this incomplete melting is what often damages the check ring on subsequent start-up.

7. Watch screw during injection: observe speed, transfer, speed after transfer, ram bounce after transfer, lack of cushion, check amount of decompression, observe screw recovery; if possible look for screw rotation during injection.

8. Stop mold during opening before any ejection and look for signs of corner lifting, etc. which might indicate part trying to stick in stationary mold half.

9. Never assume water flow is present; verify via flow indicators. Even flow into a bucket can be misleading if a low ΔP exists between supply and return lines (checking water flow into bucket will not be pushing into normal back pressure on return line circuit). A large ΔT is usually a sure sign of low flow rates.

10. Check barrel heater band functionality often; new machines can be purchased with burnout alarms.

11. Always purge heat sensitive resins after idle time; especially on molds with tendency to stick.

12. Use as large of nozzle orifice as possible, but slightly smaller than sprue orifice.

13. Prior to shut down of a leaking hot runner system, change colors of resin to indicate location of leak (can feed several handfuls new color through magnet or feed throat); inject a few shots into mold after color change through nozzle is verified by air shots.

14. Excessive clamp tonnage on undersized molds can result in platen warp/wrap causing mold to flash.

15. Perform balance of fill analysis.

16. Determine gate seal time as a guide to set injection forward time.

17. Always establish and record the transfer weight and time if using decoupled molding.

18. When purging hygroscopic resins, observe air shot purgings for foaming; this indicates presence of moisture (assuming reasonable melt temperature).

Tips, Tricks & Shortcuts (ASCII & Symbol Codes)

19. Always cool clamping plates to prevent heating of platens; heated platens result in thermal expansion which results in tie bar movement effecting the clamp close and mold protection pressures required.

20. Never mix acetal and PVC ... HCI from over heated PVC + formaldehyde may result in an explosion.

21. Methyl gas formed by degraded polypropylene may attack and erode copper alloys.

22. Locate mold half alignment blocks on horizontal and vertical center lines; use parallel locks when vertical shutoffs exist (or near vertical). Tapered interlocks will not perform full alignment until mold is closed.

23. Color code water lines (i.e. blue "ins" and red "outs").

24. Remember the tangent value for an angle of 1 degree which is 0.01745. Can then multiply distance in inches x angle x 0.01745 = rise.

Useful ASCII code symbols (press ALT and at same time type numbers from numerical keypad on right side of keyboard) and use Symbol fonts.

Arial font (and others)		Symbol fonts	
ALT+	Character	typed	displayed
155	¢	s	σ
171	½	S	Σ
172	¼	p	π
174	«	D	Δ
175	»	r	ρ
230	µ	b	β
241	±	m	µ
246	÷	a	α
248	°	d	δ
249	•	f	φ
250	·	h	η
253	^{2}	\	∴
		q	θ
		F	Φ
		W	Ω
		@	≅
		^	⊥

Calculate Heat Load and Cooling Required

$$\frac{\text{Heat Load (resin thruput x spec heat (resin) x }\Delta T)}{\text{spec heat (coolant) x temp rise of coolant}} = \text{lbs / hr of coolant required}$$

$$\frac{\dfrac{\text{lbs}}{\text{hr}}\text{ of coolant required}}{\text{coolant }\dfrac{\text{lbs}}{\text{gal}}\text{ x }60\dfrac{\text{min}}{\text{hr}}} = \text{GPM of coolant required}$$

EXAMPLE CONDITIONS (see calculation below)
shot wt = 221.5 grams (0.487 Lbs)
cycle time = 15 sec
resin = PP copolymer
spec heat PP = 0.64 Btu/(Lb °F)
spec heat water = 1.003 Btu/(Lb °F) ... (@ 55°F)
ejected part temp = 120 °F
melt temp = 440 °F
ΔT = 320 °F
temp rise in water = 5 °F (this is max allowable temp rise; 2-3° F is suggested)
1 gal water = 8.33 lbs ... 1 Ton chiller = 12,000 Btu/Hr

$$\frac{0.487 \text{ lbs}}{15 \text{ sec}} \text{ x } \frac{3600 \text{ sec}}{\text{hr}} = 116.88 \frac{\text{lbs}}{\text{hr}}$$

$$116.88 \frac{\text{lbs}}{\text{hr}} \text{ x } \frac{0.64 \text{ Btu}}{\text{lb °F}} \text{x } 320 \text{ °F } (\Delta T) = 23,937 \frac{\text{Btu}}{\text{hr}}$$

$$\frac{23,937 \dfrac{\text{Btu}}{\text{hr}}}{1.003 \dfrac{\text{Btu}}{\text{lb °F}}\text{x } 5 \text{ °F}} = 4,773 \frac{\text{lbs}}{\text{hr}} \text{ of coolant}$$

$$\frac{4,773 \dfrac{\text{lbs}}{\text{hr}}}{8.33 \dfrac{\text{lbs}}{\text{gal}}\text{x } 60\dfrac{\text{min}}{\text{hr}}} = 9.54 \text{ GPM of coolant reqd}$$

$$\frac{23,937}{12,000} = 2 \text{ tons of chiller capacity required}$$

NOTE: If all lines checked collectively together, their is risk that the GPM may not be present where heat load is present (specifically at core/cavity cooling circuits). This relates to disadvantage in connecting too many INS & OUTS if sufficient water is not available to keep them all flowing at high rate. Flow indicator/regulators work well at indicating free flowing channels where little work is done; these can be throttled down to increase flow thru channels with higher heat loads. A high percentage of available water will take the path of least resistance thru free flowing channels (e.g. clamp plates and manifold plates).

Calculating Reynolds Number

It is important for the coolant flowing thru a mold to be in turbulent flow so that the coolant nearest the O.D. of the water line doesn't do all the work; turbulent flow allows more of the water to become exposed to the core/cavity coolant channels. Reynolds number is the number which indicates if and when the flow is turbulent. A Reynolds number (Nr) of 10,000 is <u>desired</u> for optimum cooling (bare minimum of 4000 required). A quick rule of thumb to verify proper Reynolds number (and turbulent flow) is having flow (GPM) equal to three times the coolant channel ID (round or equivalent round). Note: Equiv hyd diam = 4A/P (a = area; p = perimeter).

$$Nr = \frac{7740VD}{n} \quad \text{or} \quad \frac{3160}{Dn}$$

V = fluid velocity in ft/sec

D = diameter of passage in inches

n = kinematic viscosity in centistokes

Q = coolant flow rate in gpm

TABLE of KINEMATIC VISCOSITY FOR WATER	
° F	VISCOSITY CENTISTOKES
32	1.79
40	1.54
50	1.31
60	1.12
70	0.98
80	0.86
90	0.76
100	0.69
120	0.56
140	0.47
160	0.40
180	0.35
200	0.31
212	0.28

NOTES:
1. The chart (above right) is for water; an ethylene/glycol mix changes the viscosity and may result in laminar flow unless very high flow rates are achieved. Ethylene/glycol has lower thermal conductivity and a lower specific heat (0.575 vs 1); thus, carrying less heat out of mold.
2. Attempting to connect all water lines as separate INS and OUTS will normally result in lower flow rates thru each individual circuit (versus some use of loops or bypass lines). These reduced flow rates may result in reduced heat transfer. Heat transfer coefficients calculated using Reynolds, Prandtl and Nusselt numbers continue to increase as Reynolds number increases (but, heat transfer assumes the heat conducts to the cooling channel wall ID as fast as the coolant can remove it). Therefore, it is suggested to use loops as needed to make mold setup easier and to keep flow rates elevated, but temperature rise in the circuit should not exceed 5° F (1–3° F is normal and suggested).
3. The Nr is valid only for the location for which it is calculated ... meaning a specific core/cav location. DO NOT calculate a Nr for a given entrance water line size if the size changes elsewhere in the mold where the work is done.
4. A water line in parallel should have its actual flow rate recalculated accordingly (if the measured flow is prior to the parallel branching).

API Schedule 40 & 80 Pipe Data

WATER FLOW IN PIPES

Water flow in a pipe can be approximated by the following equation, but a velocity of 5-8 ft/sec is often used instead for planning purposes (e.g. take velocity such as 7 ft/sec X CSA to get ft³/sec for conversion to GPM).

$$V = C \times \sqrt{\frac{hD}{L + 54D}}$$

V= mean velocity (ft/sec)
C= coefficient (see table below)
D= diameter of pipe (ft)
h= total head (ft) ... (1 psi = 2.309 ft head)
L= total length of pipe line (ft)

PIPE SIZE (IN)	SCHD NO.	ID (IN)	CSA (FT²)	OD (IN)	WALL (IN)	COEFF
1/8	40	0.269	0.0004	0.405	0.068	
1/4	40	0.364	0.0007	0.540	0.088	
3/8	40	0.493	0.0013	0.675	0.091	
1/2	40	0.622	0.0021	0.840	0.109	
3/4	40	0.824	0.0037	1.050	0.113	
1	40	1.049	0.0060	1.315	0.133	20.00
1	80	0.957	0.0050	1.315	0.179	19.5
1-1/4	40	1.380	0.0104	1.660	0.140	24.05
1-1/4	80	1.278	0.0089	1.660	0.191	23.46
1-1/2	40	1.610	0.0141	1.900	0.145	25.39
1-1/2	80	1.500	0.0123	1.900	0.200	24.75
2	40	2.067	0.0233	2.375	0.154	28.06
2	80	1.939	0.0205	2.375	0.218	27.31
2-1/2	40	2.469	0.0332	2.875	0.203	30.23
3	40	3.068	0.0513	3.500	0.216	32.23
3	80	2.900	0.0459	3.500	0.300	31.67
4	40	4.026	0.0884	4.500	0.237	35.07
4	80	3.826	0.0798	4.500	0.337	34.57
6	40	6.065	0.2006	6.625	0.280	39.16
6	80	5.761	0.1810	6.625	0.432	38.60
8	40	7.981	0.3474	8.625	0.322	43.30
8	80	7.625	0.3171	8.625	0.500	42.71

(APPLIES TO STEEL AND PVC)

Once velocity in ft/sec is known: multiply it by CSA in ft² to get ft³/sec; multiply ft³/sec X 7.48 X 60 for GPM.

Head losses should be added to "L" as follows:

90° Elbow has an L value equal to 1.333D

Valves have L value equal to 6D

Approximate HP REQD = (PSI X GPM)/1714 X 0.70

NOTE: Refer to specific pump curves from supplier to more accurately predict GPM & PSI output from a given pump and impeller.

COOLING TOWER REQUIREMENTS	
Air compressor	0.2 ton/HP
Hot runner system	0.25 ton/KW
Mach hyd heat exchanger	0.1 ton HP (1 HP = 0.746 KW)
Barrel cooling	1 ton/inch screw diam
Feedthroat cooling	0.5 ton/press
Plan tower pump to have 3 GPM/TON tower capacity	

Basic Algebra

COMMUTATIVE LAW FOR ADDITION & MULTIPLICATION: numbers can be added or multiplied in any order: a+b = b+a AND ab = ba ....3 x 7 = 7 x 3

ASSOCIATIVE LAW FOR ADDITION & MULTIPLICATION: the sum or product of three or more terms is unaffected by the grouping of the terms: a+b+c = a+(b+c) = (a+b)+c AND abc = a(bc) = (ab)c = (ac)b ......3 x 7 x 4 = (3 x 7) x 4

In the aforementioned, 3, 7, a, b, c, 3a, 3ab, etc are known as monomials which means the expression of a single term. Polynomials are expressions of the sum of two or more monomials such as 3a + b.

DISTRIBUTIVE LAW FOR MULTIPLICATION: When multiplying a polynomial by a monomial, each term of the polynomial is multiplied by the monomial: a(b+c) = ab + ac ......3(7 + a) = (3 x 7) + (3 x a)

OPERATIONS WITH ZERO OR NEGATIVE NUMBERS:
Note: Division by zero is impossible

$$a + (-a) = 0 \qquad\qquad -(-a) = a$$
$$a \times 0 = 0 \qquad\qquad a(-b) = -ab$$
$$0/a = 0 \qquad\qquad 0 \div 3 = 0$$
$$(-a)(-b) = ab \qquad\qquad (-a)^2 = a^2$$

RULES FOR ORDER OF OPERATIONS:
1. Perform operations in parentheses first (if any).
2. Perform exponential functions and/or roots second.
3. Perform multiplication and division from left to right.
4. Perform addition and subtraction from left to right.

Example – Order of Operations:
Note: a x b; a X b; a•b; ab; a(b); (b)a ....are ways to write: a times b.

Example 1:

$$3 \times (7 + 4) + 3 \times 8^2 = Y$$
$$3 \times 11 + 3 \times 8^2 = Y$$
$$33 + 3 \times 64 = Y$$
$$33 + 192 = Y$$
$$225 = Y$$

Example 2:

$$3 \times \sqrt{(7 + 4^2)} + 3 \times 8^2 = Y$$
$$3 \times \sqrt{(7 + 16)} + 3 \times 64 = Y$$
$$3 \times \sqrt{23} + 192 = Y$$
$$3 \times 4.796 + 192 = Y$$
$$14.388 + 192 = Y$$
$$206.39 = Y$$

Basic Algebra (exponents, fractions)

RULES FOR POWERS AND EXPONENTS:

$$a^m \times a^n = a^{m+n}$$

$$\frac{a^m}{a^n} = a^{m-n} \qquad (ab)^n = a^n b^n$$

$$a^{-n} = \left[\frac{1}{a}\right]^n = \frac{1}{a^n}$$

$$(a^m)^n = a^{mn} \qquad a^0 = 1; \; 0^n = 0$$

RULES FOR FRACTIONS:

$$\frac{a}{c} \pm \frac{b}{d} = \frac{ad \pm bc}{cd}$$

$$\frac{a}{c} \pm \frac{b}{c} = \frac{a \pm b}{c}$$

$$\frac{a}{c} \pm \frac{a}{d} = \frac{a(d \pm c)}{cd}$$

$$\frac{a}{b} \times \frac{c}{d} = \frac{ac}{bd}$$

$$\frac{\frac{a}{b}}{\frac{c}{d}} = \frac{a}{b} \times \frac{d}{c} = \frac{ad}{bc}$$

Note:
These rules for dividing fractions must be understood. When you divide one fraction by another; you can multiply by the reciprocal. Dividing fractions will be seen often in the course of converting numbers and units.

In the example to the right, the rules above are demonstrated with the following values:
a = 3, b = 2, c = 8, d = 5

$$\frac{3}{8} + \frac{2}{5} = \frac{3 \times 5 + 2 \times 8}{8 \times 5} = \frac{15 + 16}{40} = \frac{31}{40}$$

$$\frac{3}{8} + \frac{2}{8} = \frac{3 + 2}{8} = \frac{5}{8}$$

$$\frac{3}{8} + \frac{3}{5} = \frac{3(5 + 8)}{8 \times 5} = \frac{3 \times 13}{40} = \frac{39}{40}$$

$$\frac{3}{2} \times \frac{8}{5} = \frac{3 \times 8}{2 \times 5} = \frac{24}{10} = 2.4$$

$$\frac{\frac{3}{2}}{\frac{8}{5}} = \frac{3}{2} \times \frac{5}{8} = \frac{3 \times 5}{2 \times 8} = \frac{15}{16}$$

Ratios

Ratios are often used to compare one quantity in terms of another quantity of the same kind. For example: One part weighs 32 grams and another weighs 8 grams; there exists a ratio between them of 32:8. 32:8 can also be expressed as 32 to 8, 32 ÷ 8 or 32/8.

RULES FOR RATIOS: Both terms of any ratio can be multiplied or divided by the same number; that is: 32:8 or 32/8 can reduce to 4:1 or 4/1 dividing both numbers by 8.

Proportions

A proportion is a statement of equality between two ratios
(3/8 is proportional to 9/24 ....and.... 3/8 = 9/24):

$$3:8 \ :: \ 9:24$$

extremes

$$3:8 \ :: \ 9:24$$

means

OR

$$\frac{3}{8} = \frac{9}{24}$$

3 & 24 are extremes

RULES FOR PROPORTIONS:
1. The product of the means equals the product of the extremes.
   3 x 24 =72 as does 8 x 9 = 72
2. The product of the means divided by either extreme gives the other extreme.
   8 x 9 =72.....72 ÷ 24 = 3 as does 72 ÷ 3 = 24
3. The product of the extremes divided by either mean gives the other mean. 3 x 24 =72.....72 ÷ 9 = 8 as does 72 ÷ 8 = 9

Knowing these rules, we can solve for unknowns as follows:

Example 1:
1000 parts weighs 45 grams; thus, our part weight is 0.045 grams each. A ratio can be established whereby 45 grams:1000 parts is equivalent to solve for an unknown number of parts.

In this example, we cross multiply to solve for unknown "Y" number of parts. Cross multiplication is a process allowed by rule #1 above under rules for proportions. We then used rule #2 whereby the product of the means divided by either extreme gives the unknown extreme.

$$\frac{45 \ gr}{1000 \ parts} = \frac{288 \ gr}{Y \ parts}$$

$$45 \ gr \times Y = 288 \ gr \times 1000 \ parts$$

$$Y = \frac{288 \ gr \times 1000 \ parts}{45 \ gr}$$

$$Y = 6400 \ parts$$

Example 2:
One person can hand assemble 1600 products for shipment each 8 hrs; there is a need to assemble 20,000 products. How many hrs are required?

$$\frac{8 \ hrs}{1600 \ assm} = \frac{Y}{20,000 \ assm}$$

$$8 \ hrs \times 20,000 \ assm = 1600 \ assm \times Y$$

$$\frac{8 \ hrs \times 20,000 \ assm}{1600 \ assm} = Y$$

$$100 \ hrs = Y$$

It will take one person 100 hrs or 12½ days to assemble 20,000 assemblies (8 hr days).

Conversion Factors

LENGTH

1 inch = 25.4 mm
1 mm = 0.03937 in
1 foot = 30.48 cm
1 micron = 0.001 mm
1 micron = 0.0000394 in
1 inch = 2.54 cm
1 meter = 39.37 in
1 meter = 100 cm
1 microinch = 0.000001 in
1 microinch = 0.0254 microns
(printer's)
1 pica = 0.166 in
1 point = 0.01384 in

WEIGHT

1 lb = 453.6 gr
1 lb = 16 oz
1 gram = 0.035 oz
1 kg = 1000 gr
1 kg = 2.2046 lb
1 oz = 28.35 gr
1 metric ton = 2204.6 lb
1 metric ton = 1000 kg

ANGLES

1 degree = 0.01745 radian
1 degree = π/180 radian

VOLUME

1 cu in = 16.387 cc
1 cu ft = 1728 cu in
1 qt = 0.946 L
1 gal = 128 oz
1 cc = 1 gr (water)
1 gal = 8.33 lb
1 cu ft = 7.48 gal

AREA

1 sq in = 6.452 cc
1 sq ft = 144 sq in
1 acre = 43560 sq ft
1 sq cm = 0.155 sq in
1 sq ft = 0.111 sq yd
1 sq mm = 0.00155 sq in
1 sq km = 0.3861 sq mi

PRESSURE

1 in Hg = 13.6 in H_2O
1 kg/cm^2 = 14.223 psi
1 bar = 14.5 psi
1 atmos = 14.696 psi
1 MPa = 145 psi

ENERGY

1 BTU = 777.97 ft lb
1 cal = 3.09 ft lb
1 BTU = 252 cal
1 kwh = 3412 BTU
1 H.P. = 746 watts
1 ton (refrig) = 12000 Btu/hr
1 ton (refrig) = 3517 watts

SPECIFIC HEAT & HEAT TRANSFER

1 Cal/sec cm °C = 2903 BTU-in/hr ft^2 °F
1 Cal/sec cm °C = 241.9 BTU/hr ft °F
1 W/(m °K) = 0.0023884 Cal/sec cm °C
1 W/(m °K) = 0.5778 BTU/hr ft °F
1 W/(m °K) = 6.9335 BTU-in/hr ft^2 °F
1 Btu/(Lb °F) = 4.184 KJ/(Kg °K)
1 Btu/(Lb °F) = 4184 J/(Kg °K)
1 Cal/(g °C) = 1 Btu/(Lb °F)

TEMPERATURE CONVERSIONS

°C = (°F-32)/1.8
°F = (°C x 1.8) + 32
°K = (°F+459.67)/1.8

Basic Conversions

Conversion factors are very easy to use. The following examples and comments should make the use of conversion factors easier to accomplish. When converting, we start with anything that we know. If you are assigned the task of packaging 3 pounds of plastic for shipment, but the scales only read in grams; we could start with what we know - 3 lbs ... 3 lbs = How many grams?

At this point, it should be pointed out that the conversion factors seen on previous page can be re-written as (see example right):

$$1 lb = 453.6\ gr = \frac{453.6\ gr}{1 lb} = \frac{1 lb}{453.6\ gr}$$

If what is on top is equivalent to what is on bottom ... anything goes.

$$1 lb = 453.6\ gr = \frac{453.6\ gr}{1 lb} = \frac{16\ oz}{1 lb}$$

Now, back to our problem: if we start with lbs and we want to end up with grams, we use a conversion factor to convert.

We will multiply 3 lbs by a conversion factor: $\frac{453.6\ gr}{1 lb}$

We can rewrite the above as: $\frac{3\ lb}{1} \times \frac{453.6\ gr}{1 lb} = ??$

Now we cancel the units: anything seen on top will cancel one set like it on the bottom: the lbs on top will be cancelled (erased) by the lbs on the bottom. Now we multiply 3 X 453.6 to get 1360.8 on the top; the bottom will be 1 X 1 to get 1. The only units left are grams and they are on top which is where they need to be if we want grams; thus, the answer is 1360.8 grams.

The main point is to identify what is known, arrange the conversion factor such that units can be cancelled and the desired units are on top. Often times, several conversion factors are needed. We can put any number in there, as long as they are valid factors whereby the top is equivalent to the bottom. See the example right whereby we want to know how many ounces of coffee are in a 5 gallon pot:

$$\frac{5\ gal}{1} \times \frac{4\ qt}{1\ gal} \times \frac{2\ pt}{1\ qt} \times \frac{2\ cup}{1\ pt} \times \frac{8\ oz}{1\ cup} = ??$$

After we do the unit cancellation and the multiplication, we get 640 oz.

Conversion factors can have numbers other than 1 on the lower half of the fraction; How many lbs in 3000 grams? After unit cancellation and the multiplication and division, we get 6.614 lbs after round off. Note: Always keep track of the units and perform all the necessary division.

$$\frac{3000\ grams}{1} \times \frac{1 lb}{453.6\ grams} = ??$$

See sample calculations below to get density & thermal conductivity converted to proper units for use in a cooling time calculation (see also pages on cooling time calculation).

$$0.90\frac{gr}{cm^3} \times \frac{16.38\ cm^3}{1\ in^3} \times \frac{1728\ in^3}{1\ ft^3} \times \frac{1 lb}{453.6\ gr} = 56.16\ \frac{lb}{ft^3}$$

$$0.1255\frac{W}{(m\ ^\circ K)} \times \frac{0.5778\frac{B\text{-}u}{hr\ ft\ ^\circ F}}{\frac{W}{(m\ ^\circ K)}} = 0.0725\ \frac{B\text{-}u}{hr\ ft\ ^\circ F}$$

Percentages (Examples 1 & 2)

The word "percent" comes from "per" – meaning "for each" and "cent" which is based on the Latin word "centum" – meaning one hundred. Percent puts numbers in perspective on a scale from one to a hundred. Percent is often used to describe change. When describing change it is important to establish a basis for comparison – this basis becomes our denominator; the change is the numerator.

Example 1:
Jim makes $12.35 per hour and gets a 55¢ per hour raise. What percent raise did Jim get?

$$\frac{change}{basis\ for\ comparison} = Decimal\ equivalent\ of\ \%\ change$$

$$\frac{(12.90 - 12.35)}{12.35} = \frac{0.55}{12.35} = 0.0445 = 4.45\%$$

$$0.0445 \times 100 = 4.45\%$$

It must now be pointed out that percentages are written as one hundred times larger than their decimal equivalent. As shown above, the math returns a value of 0.0445; we multiply times 100 to get percent; thus, 4.45%. Recall that "percent" means "per hundred"; thus, 4.45% is same as 4.45/100 which equals 0.0445.

RULES FOR PERCENT: When we change a decimal to a percent: we multiply by 100 which moves the decimal point two places to the right. The number 0.0445 equals 4.45%. When we change a percent to a decimal: we divide by 100 which moves the decimal two places to the left. The percent 4.45% becomes 0.0445. We can then say 0.0445 is the decimal equivalent to 4.45%.

We can now say that Jim makes 4.45% more than he did. When we describe percentages, we must make math agree with the statement.

Example 2:
Jim works to improve the overall molding cycle on a press and succeeds in reducing the cycle from 29 seconds to 25.5 seconds.

$$\frac{(29 - 25.5)}{29} = \frac{3.5}{29} = 0.1207 = 12.07\%$$

Jim has accomplished a 12.07% cycle reduction. The new cycle is 12.07% faster than old cycle. If we later slow back down to 29 seconds, the percent cycle time increase will be 14% because the basis for comparison will have changed. The denominator is the basis for comparison; thus, our new answer for percentage increase is 0.14 or 14%.

$$\frac{(29 - 25.5)}{25} = \frac{3.5}{25} = 0.14 = 14\%$$

Percents are useful to put things in perspective as mentioned above. If we change the cycle on a different molding job by 3.5 seconds and it was running at 11 seconds before the change; this yields a 31.8% cycle reduction; thus, 3.5 seconds is more significant in different situations and better described as a percent change.

Percentages (Examples 3, 4, 5 & 6)

Percentages are also used in ways different from the aforementioned discussion. When calculating efficiency, we compare actual performance to a plan. We want to express this efficiency relative to 100% efficient – no better or no worse than expected. Numbers higher than 100 are favorable and numbers lower than 100 are unfavorable.

Examples 3, 4, 5 & 6:

Jim calculates his labor utilization by dividing his total labor hours each shift by what is estimated to be required. This estimated requirement may be different for each job; thus, there is a sum total of labor to be used each day to run the scheduled production. Jim has 17 presses running which are estimated to require 7 people (3 inspector-packers, 2 technicians; 1 material handler and 1 stock chaser for finished goods). Jim hires 2 extra temporary workers to help sort defective product; thus, Jim is using 9 people. What is Jim's labor utilization?

$$\frac{plan}{actual} = \frac{7\ hrs}{9\ hrs} = 0.7777 = 77.77\%$$

Jim also calculates his cycle efficiency in a similar manner whereby each job is to run a prescribed cycle based on original estimates. One job is estimated to run 29 seconds, but Jim has reduced this cycle to 25.5 seconds. Jim's cycle efficiency for this job is as follows:

$$\frac{plan}{actual} = \frac{29sec}{25.5sec} = 1.1372 = 113.72\%$$

Jim later learns that the scrap for this job with the reduced cycle was reported by the customer to be 12%. Jim knows that he produced 50,176 parts that were shipped using that cycle. How many defective parts does the 12% represent? We can calculate this by using our same basic equation, then multiply both sides of the equation by 50,176 to solve for the unknown "X" – defective parts:

$$\frac{defective}{basis} = \frac{X}{50,176} = 0.12$$

$$\frac{\cancel{50,176}}{1} \times \frac{X}{\cancel{50,176}} = 0.12 \times 50,176$$

NOTE: 0.12 is decimal equiv. of 12%; we need to use the decimal equiv. in our calculation

$$X = 6021\ defective\ parts$$

RULES FOR PERCENT: When we multiply or divide percents; we must convert percent back to it's decimal equivalent before use.

Jim's company suffers in stockholder confidence whereby the stock drops from $14.15 per share to $11.45 per share; the percent drop is as follows:

$$\frac{(14.15 - 11.45)}{14.15} = \frac{2.70}{14.15} = 0.1908 = 19.08\%$$

In this example, one must decide what is the basis for comparison. Since we want to know the percentage drop, we use $14.15 – where we came from. If the stock later rises $2.70, then the percent increase back to $14.15 per share will be 23.58% increase.

A Method For Estimating The Cooling Cycle

$$Q = \frac{-(t^2)}{2\,\pi\alpha} \times LN \left[\frac{\pi\,(T_x - T_m)}{4\,(T_c - T_m)} \right]$$

See list below for explanations of terms; see table farther below for values.

Q = cooling time (sec) ... t = part thickness (inches)

α = thermal diffusivity (in²/sec) = K ÷ ρ Cp

K = thermal conductivity (Btu/hr ft °F) ... ρ = density (lb/ft³)

Cp = specific heat (Btu/lb°F)

Tx = ejected part temperature; use heat deflection temp @ 66 psi (use table less 30°)

Tm = mold temperature (°F) ... Tc = cylinder or melt temperature (°F)

LN = natural logarithm

NOTE: Cooling time using this formula becomes somewhat conservative or long for thicker part walls. This is because actual moldings of thick walled parts typically result in opening the mold prior to full wall cooling (a hotter or slightly molten center may exist on thick walled parts). Post mold part handling is critical in such scenarios.

RESIN (abbr)	Thermal conductivity K[1] (Btu/hr ft °F)	Cp[1] (Btu/lb°F)	Density[1] (gr/cc)	Density[1] lb/ft³	Thermal diffusivity[1] (in²/sec)	Deflection Temp[1] @ 66 psi
ABS	0.108 - 0.192	0.490	1.060	66.144	0.000185	203
CA, CAP	0.100 - 0.192	0.410	1.260	78.624	0.000181	192
CAB	0.100 - 0.192	0.390	1.200	74.88	0.000200	201
HIPS	0.024 - 0.073	0.501	1.050	65.52	0.000059	185
IONOM	0.142 - 0.142	0.645	0.950	59.28	0.000148	125
LDPE	0.192 - 0.192	0.760	0.920	57.408	0.000176	113
MDPE	0.192 - 0.242	0.765	0.935	58.344	0.000194	155
HDPE	0.267 - 0.300	0.870	0.960	59.904	0.000217	186
PA6 GF	0.173 - 0.173	0.669	1.380	86.112	0.000120	460
PA 6, 6/6	0.142 - 0.142	0.731	1.140	71.136	0.000109	356
PC	0.108 - 0.108	0.438	1.200	74.88	0.000132	280
PET	0.144 - 0.144	0.502	1.330	82.992	0.000138	153
PET (C)	0.144 - 0.144	0.548	1.360	84.864	0.000124	252
PMMA	0.092 - 0.142	0.454	1.190	74.256	0.000138	215
POM	0.133 - 0.133	0.715	1.420	88.608	0.000084	336
PP	0.066 - 0.079	0.667	0.900	56.16	0.000077	204
PP co	0.048 - 0.100	0.640	0.900	56.16	0.000083	184
PPO/PS	0.125 - 0.125	0.519	1.070	66.768	0.000144	234
GF PPS	0.167 - 0.259	0.497	1.650	102.960	0.000166	500
PS g.p.	0.058 - 0.080	0.480	1.060	66.144	0.000087	180
PSU	0.150 - 0.150	0.520	1.240	77.376	0.000149	345
PVC	0.083 - 0.083	0.383	1.290	80.496	0.000107	156
PVC rig	0.092 - 0.092	0.340	1.400	87.36	0.000123	174
SAN	0.070 - 0.070	0.471	1.080	67.392	0.000088	225

THERMAL PROPERTIES FOR SELECTED RESINS

[1] Increased accuracy will result if actual data for your specific resin is obtained from your resin supplier due to variation between different blends from different suppliers. Thermal diffusivity is calculated from density (lb/ft³), specific heat (Btu/lb°F) & thermal conductivity (Btu/hr ft °F). Note: These aforementioned units must be used together for proper unit cancellation to get correct answer. You must convert metric units or use only metric units for proper unit cancellation.

Cooling Time Math: Cooling Time Example 1

In the example below we calculate cooling time based on the following conditions:

Polycarbonate part that is 0.045 inches thick (shot weight is 85 grams). Process that includes a 580° F melt temp & 180° F mold temperature and a fill time of 0.35 seconds.

In this example, we will use data from the table on the preceding page: thermal diffusivity equals 0.000132 in²/sec and our ejected part temperature will be 250° F (Tx).

$$Q = \frac{-(t^2)}{2\pi\alpha} \times LN\left[\frac{\pi}{4}\frac{(T_x-T_m)}{(T_c-T_m)}\right]$$

$$Q = \frac{-(0.045\ in^2)}{2\times3.1416\times0.000132\ \frac{in^2}{sec}} \times LN\left[\frac{\pi}{4}\times\frac{(250-180\ °F)}{(580-180\ °F)}\right]$$

$$Q = \frac{-(0.002025\ in^2)}{0.000829\ \frac{in^2}{sec}} \times LN\left[\frac{3.1416}{4}\times\frac{70\ °F}{400\ °F}\right]$$

$Q = -2.4427\ sec \times LN\ 0.137445$

$Q = -2.4427\ sec \times -1.9845$

$Q = 4.85\ seconds\ cooling\ time\ req'd$

Planned Process

0.35 fill time

2.00 pack (& cool)

1.85 plasticize (& cool)

1.00 additional cool

Note: There is cooling during pack, plasticize and cool

0.95 open

1.95 robot traverse & part ejection

1.40 close and build tonnage

9.50 second total cycle time

The 4.3 sec total open time is 45.3% of cycle!

$$\frac{4.3}{9.5} = 0.453 \times 100 = 45.3\%$$

Calculation Notes:

1. A negative times a negative equals a positive.
2. Since seconds were on the bottom of bottom; it goes on top; remember that division by a fraction is same as multiplying by the reciprocal.
3. The calculation in brackets is performed; then a single answer is typed into a scientific calculator (calculator w/ LN function), then simply press LN button and read display. LN stands for natural logarithm (previously known as log_e).
4. The thickness is squared inside of brackets with minus sign outside to preserve the negative value of squared product. We need a negative to cancel the negative created by the LN operation.
5. As can be seen the units cancel fine when units from table on proceeding page are used. If other units arc used they should be converted as needed to cancel. Metric units can be used if converted to units which will cancel. See next page for additional practice with unit cancellation or conversion.

Cooling Time & Converting Units

In the example below we calculate cooling time based on the following conditions (Note the unit cancellations - shown with different angled hash marks for this example).

Polypropylene part that is 0.045 inches thick.

Process that includes a 425° F melt temp & 75° F mold temp plus a fill time of 0.70 seconds and total mold open time of 3.66.

Resin supplier provides a density of 0.90 gr/cm³, a thermal conductivity of 0.1255 W/m°K, a specific heat of 2678 J/Kg°K and an HDT (heat deflection temperature) of 212 °F ... we subtracted 30 to get a Tx of 182 °F for use on next page ... (get specific heat & heat transfer conversion factors from page 88). If we do not have better supplier data, then use table on page 92 where the Tx for PP is listed as 204 (204 - 30 = 174).

#1 Calculation to convert density to needed units:

$$0.90\frac{gr}{cm^3} \times \frac{16.38\ cm^3}{1\ in^3} \times \frac{1728\ in^3}{1\ ft^3} \times \frac{1Lb}{453.6\ gr} = 56.16\frac{Lb}{ft^3}$$

#2 Calculation to convert thermal conductivity to needed units (0.5778 from p. 88):

$$0.1255\frac{W}{(m \times °K)} \times \frac{0.5778\ \dfrac{Btu}{hr \times ft \times °F}}{\dfrac{W}{(m \times °K)}} = 0.0725\frac{Btu}{hr \times ft \times °F}$$

#3 Calculation to convert specific heat to needed units (4184 from p. 88):

$$2678\frac{J}{(Kg \times °K)} \times \frac{1\ \dfrac{Btu}{Lb \times °F}}{4184\ \dfrac{J}{(Kg \times °K)}} = 0.640\frac{Btu}{Lb \times °F}$$

#4 Calculation of thermal diffusivity using the needed units to yield in²/sec so units will properly cancel in cool time equation on opposite page:

$$\alpha = \frac{K}{\rho \times C_p}$$

$$\alpha = \frac{0.0725\ \dfrac{Btu}{hr \times ft \times °F} \times \dfrac{144 in^2}{1 ft^2}}{56.16\dfrac{Lb}{ft^3} \times 0.640\dfrac{Btu}{Lb \times °F} \times \dfrac{3600 sec}{1 hr}}$$

$$\alpha = 0.000081\ {in^2}/{sec}$$

Calculating Cooling Time: Cooling Time Example 2

#5 Using the aforementioned, we can calculate the cooling time. Note: The 182 came from supplier listed HDT of 212 °F ... subtract 30 from HDT as stated on page 92 for Tx). The Tc is Melt or Cylinder/Barrel temperature given as 425 °F & Tm is mold temperature given as 75 °F ... we solved for a thermal diffusivity on previous page and the wall thickness was given as 0.045 inches:

$$Q = \frac{-(t^2)}{2\,\pi\,\alpha} \times LN\left[\frac{\pi}{4} \times \frac{(Tx-Tm)}{(Tc-Tm)}\right]$$

$$Q = \frac{-(0.045\ \text{In}^2)}{2 \times 3.1416 \times 0.000081\ \text{In}^2/\text{sec}} \times LN\left[\frac{\pi}{4} \times \frac{(182°F-75°F)}{(425°F-75°F)}\right]$$

$$Q = \frac{-0.002025\ \text{sec}}{0.0005089} \times LN\left[\frac{3.1416}{4} \times \frac{107°F}{350°F}\right]$$

$$Q = \frac{-0.002025\ \text{sec}}{0.0005089} \times LN\,(0.240108)$$

$$Q = -3.97886\ \text{sec} \times -1.42666$$

$$Q = 5.68\ \text{seconds}$$

If the cooling time is 5.68 seconds and the fill time is 0.70 seconds and the total mold open time is 3.66 seconds ... then the overall cycle would be 10.04 seconds.

Note: This formula predicts an excessively long cooling time for thicker walls because it assumes the entire wall will be cooled to the Tx temperature from the Tc temperature ... in practice, we likely open mold and eject thicker parts with only a large portion of wall thickness cooled to the Tx. At thicker wall thicknesses, the equation may (will) predict too long of a cool time. If you use this equation instead of other more sophisticated cycle estimator programs ... you may want to experiment with factoring wall thickness downward (e.g. if wall thickness is greater than 0.080 inches than use 80% of wall thickness in equation to better simulate what likely would occur on molding floor - eject part with a hotter wall core temperature ... food for thought).

Interpolation

Sometimes it is necessary to lookup numbers from tables to use in various calculations. These tables typically list only a sampling of values; and thus, may not provide the exact lookup value that you are seeking. A case in point is the table (right) for specific heat, but the table could be for thermal conductivity, thermal expansion, etc.

Specific Heat of Water	
Temp °F	Cp Btu/(lb °F)
32	1.007
50	1.003
100	0.998
150	1.000
200	1.006
212	1.008
250	1.015
300	1.029
350	1.060
400	1.085
450	1.116
500	1.180

(150 / 200: 180 = ?)

If we want to calculate water flow required, we will use the formula below. Using this formula, we will determine the specific heat of water at 180° F. We can estimate this to be around 1.004, but we want to know for sure. In the coolant flow calculation, the estimated 1.004 would likely be sufficient, but for purposes of instruction we will calculate precisely the specific heat for 180° F. Some tables might have a larger spread between values whereby estimating the needed value would result in increased error.

Formula for Coolant Flow Req'd in an Injection Mold

$$\frac{\text{Heat Load (Resin thruput X spec heat (resin) X } \Delta T)}{\text{spec heat (coolant) X temp rise of coolant}} = \text{lbs/hr of coolant reqd}$$

$$\frac{\frac{\text{lbs}}{\text{hr}} \text{ of coolant required}}{\text{coolant } \frac{\text{lbs}}{\text{gal}} \text{ X } 60 \frac{\text{min}}{\text{hr}}} = \text{GPM of coolant required}$$

Use Interpolation ... Specific Heat of Water at 180°F

$$\frac{180 - 150}{200 - 150} = \frac{30}{50} = 0.60 \text{ (or 60 %)}$$

$$0.60 \times (1.006 - 1.000) + 1.000 = \text{spec heat @ 180°F}$$

$$0.60 \times 0.006 + 1.000 = \text{spec heat @ 180°F}$$

$$0.0036 + 1.000 = \text{spec heat @ 180°F}$$

$$1.0036 = \text{spec heat @ 180°F}$$

First we must assume that within each interval such as 150° to 200° there was a straight line relationship meaning even spacing of the specific heat values. As can be seen, there is not a straight line relationship from 32° to 500° because the numbers go from high to low back to high again; thus, there is "curvature", but within each interval we are forced to assume a straight line relationship.

We then make the position of 180°, a percentage of that respective range. In percentages, we relate things to a scale of 1 to 100; thus, we describe 180's position between 150 and 200 as 60% of the way between 150 and 200. Knowing this relationship of 60% above the minimum value for the range (100), we can then take 60% times the related range for specific heat ... 60% times (1.006 - 1.000) is 0.0036.

When we add 0.0036 to 1.000, we get 1.0036 as the specific heat value that relates to 180°F. 1.0036 is 60% of the way from 1.000 to 1.006. We will now use 1.0036 in the formula above for coolant flow required; see next page for this calculation.

Heat Load & Coolant Flow Required

Formula for Coolant Flow Req'd in an Injection Mold

$$\frac{\text{Heat Load (Resin thruput X spec heat (resin) X } \Delta T)}{\text{spec heat (coolant) X temp rise of coolant}} = \text{lbs/hr of coolant reqd}$$

$$\frac{\frac{\text{lbs}}{\text{hr}} \text{ of coolant required}}{\text{coolant } \frac{\text{lbs}}{\text{gal}} \text{ X } 60 \frac{\text{min}}{\text{hr}}} = \text{GPM of coolant required}$$

On a previous page (p. 93), we calculated cooling time at 4.85 sec for a polycarbonate part, the overall cycle was listed at 9.5 seconds and shot weight was listed at 85 grams (0.187 lbs). We look up the specific heat for polycarbonate to be 0.438 Btu/(lb °F) – see page 92 for table. To calculate the heat load for the mold, we first convert the resin thruput to lbs per hour.

$$\frac{0.187 \text{ lbs}}{9.5 \text{ sec}} \times \frac{3600 \text{ sec}}{\text{hr}} = 70.863 \frac{\text{lbs}}{\text{hr}}$$

We now multiply the resin thruput in lbs per hour times the specific heat of the resin.

$$70.863 \frac{\text{lbs}}{\text{hr}} \times 0.438 \frac{\text{Btu}}{(\text{lb °F})} = 31.038 \frac{\text{Btu}}{\text{hr °F}}$$

We multiply this value times the temperature change to get the total Btu(s) per hour created for which we need the coolant to remove from the mold (recall that the melt temp is 580°F and eject temp is 250°F).

$$31.038 \frac{\text{Btu}}{\text{hr °F}} \times (580°F - 250°F) = 10242.54 \frac{\text{Btu}}{\text{hr}}$$

Now we use our previously, interpolated value for the specific heat of 180 °F water which was found to be 1.0036 Btu/(lb °F). We must also decide on a maximum temperature rise for the coolant flow thru the mold ... we will use 3° F as an acceptable temperature rise.

$$\frac{10242.54 \frac{\text{Btu}}{\text{hr}}}{1.0036 \frac{\text{Btu}}{(\text{lb°F})} \times 3°F} = 3401.93 \frac{\text{lbs}}{\text{hr}} \text{ of coolant flow}$$

We now want to convert the lbs per hour coolant to GPM (gallons per minute) because we have flow meters to measure in GPM.

This minimum value assumes the heat load and coolant flow are evenly distributed, but they are not; thus, we will arbitrarily plan to have twice

$$\frac{3401.93 \frac{\text{lbs}}{\text{hr}}}{8.33 \frac{\text{lbs}}{\text{gal}} \times 60 \frac{\text{min}}{\text{hr}}} = 6.81 \frac{\text{gal}}{\text{min}} \text{ coolant flow required}$$

this value present. It also assumes that the entire cycle is available for heat transfer; this is debatable since there is coolant flow thru mold at all times, but the parts come and go ... there is some time when the mold is empty; thus, the actual overall cycle used may need to be adjusted downward to reflect the best estimate of actual cooling time. Note however, after part ejection there is residual heat in the mold still conducting toward coolant channels to be removed.

Calculating Thermal Expansion

This formula is used to calculate thermal expansion: $\delta_t = \alpha\,(\Delta T)\,L$

Where the symbols mean the following: δ_t is elongation due to temp change, α is the coefficient of thermal expansion, ΔT is temperature change ($T_{HIGH} - T_{LOW}$) & L is the length subjected to the expansion.

Example: The leader pins on a mold are 18.75 inches apart. The "A" half is heated from 70° F to 185° F and to 165° F on the ejector half; if moldbase is steel, use 6.6 X10^{-6} or 6.6 X 0.000001 as the coefficient of expansion or use table if specific steel is known.

The expansion between leader pin centers (in ejector half) is calculated as follows:

δ_t = 6.6 X 0.000001 X (165 - 70) X 18.75
δ_t = 0.0000066 X 95 X 18.75
δ_t = 0.0117 inches

The bushing holes in "A" half will expand even more at 0.01423 inches since they are in hotter plates (185° instead of 165° F). This yields a 0.0025 inch differential in leader pin to bushing spacing ... this is about the maximum that can be tolerated unless the bushing holes are worn.

Note also that the shut height of this mold is 11.751 inches. This will grow to be 11.759 inches (0.008 inches increase).

$\delta_$ = 6.6 x10^{-6} x (185 -70) x 4.563 + 6.6 X10^{-6} x (165 -70) X 7.188
$\delta_$ = 0.0000066 x 115 x 4.563 + 0.0000066 X 95 X 7.188
$\delta_$ = 0.003463 + 0.004507 inches = 0.008 inches.

This increase may be enough to throw off the mold protection setup if set very close; thus, requiring the clamp lockup position to be readjusted after temperature equilibrium is established.

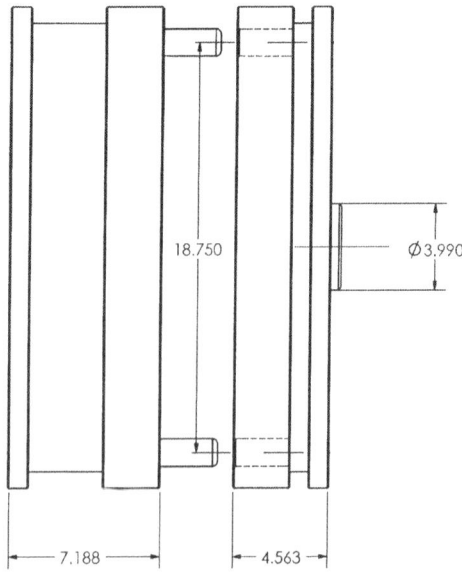

Thermal Expansion Coefficients - Mold Materials

μ = micro....multiply X .000001; μ in/in from 68 °F to

TYPE	200 °F	400 °F	800 °F
1020	6.5	6.7	7.1
4140	6.8	7.1	
6150	6.8	7.1	7.4
W1	5.8	6.1	7.3
W2	7.4		8.0
S1	6.9	7.0	7.5
S5	6.4		7.0
S6	6.4		7.0
S7	6.8	7.0	7.4
O1	5.8	5.9	7.1
O2	6.2	7.0	7.7
A2	5.8	5.9	7.2
A6	6.4	6.9	7.5
D2	5.6	5.7	6.6
D3	6.3	6.5	7.2
D4	6.2		6.9
H10	6.1		6.8
H11	6.2	6.9	7.1
H13	5.8	6.4	6.8
H14	6.1		
H19	6.1	6.1	6.7
H21	6.9	7.0	7.2
H22	6.1		6.4
T1	5.3	5.4	6.2
T5	6.2		
T15		5.5	6.1
M1		5.9	6.3
M2	5.6	5.2	6.2
M3			6.4
M4			6.4
M7		5.3	6.4
M10			6.1
L2	7.4		8.0
L6	6.3	7.0	7.0
P2	7.0		7.6
P20	6.5		7.1
SS 303,4	9.6	9.9	
SS 316	8.8	9.0	
SS 414	5.8	6.1	
SS 420	5.7	6.0	
SS 440	5.7		
Ti alpha alloy	4.6		4.8
Ti alpha-beta	5.0	5.1	5.2
Ti beta alloy	5.2	5.4	5.6
Ampco 18, 21 ??	9.0		
Ampco 940	9.7		
FREE MACH Cu	9.8		
BeCu (2%)	9.7		
BeCu (0.5%)	9.8		
6061 Aluminum	13.1		
INCONEL	6.4		

Intensification Ratio: Injection Cylinder

Intensification ratio is the mechanical advantage of the injection cylinder whereby the hydraulic oil pressure acting on a large surface area creates a pressure multiplier that causes plastic pressure entering the mold to be 8 - 30 times greater (10 -16 is most common). Typical system hydraulic pressures are approximately 2,000 psi; thus, plastic pressures would become 30,000 psi if the intensification ratio was 15:1.

There are two ways to effectively determine the machine's intensification ratio (based on following equation):

P1 X A1 = P2 X A2 whereby

> P1 = Plastic pressure leaving machine nozzle
> A1 = Screw area
> P2 = Hydraulic oil injection pressure
> A2 = Injection cylinder area

1) Read OEM supplied machine specification sheet for your specific machine (same screw size and injection unit type). Find the maximum injection pressure and max hydraulic pressure. The intensification ratio is simply the injection pressure (plastic) divided by the hydraulic pressure – this is typically the easiest method.

$$\text{intensification ratio} = \frac{\text{maximum injection pressure}}{\text{maximum hydraulic pressure}}$$

2) The second method requires that the screw diameter and resultant area be known as well as the injection cylinder area (total - some machines have two injection cylinders in parallel). This information will likely require the review of hydraulic schematics to identify the injection cylinder diameters).

$$\text{intensification ratio} = \frac{\text{total injection cylinder area}}{\text{screw area (based on screw diameter)}}$$

Using method #2 for barrel seen on next page (screw/barrel diameter of 50 mm and an injection cylinder ram diameter of 7.5 inches), we calculate the intensification ratio as follows:

$$50 \text{ mm} \times \frac{1 \text{ in}}{25.4 \text{ mm}} = 1.968 \text{ inches (screw diameter)}$$

$$\text{area of a round} = \frac{\pi d^2}{4} \text{ or } \pi r^2 (\pi = 3.1416)$$

$$\frac{3.1416 \times 1.968^2}{4} = 3.042 \text{ in}^2 (\text{area of screw})$$

$$\frac{3.1416 \times 7.5^2}{4} = 44.1787 \text{ in}^2 (\text{area of inj cylinder})$$

$$\frac{44.1787}{3.042} = 14.52.......\text{also written as } 14.52:1$$

Clamp Tonnage & Cavity Pressure

This same principle of pressure amplification also acts on the mold cavities. If we are running the screw and injection unit seen on previous page at 750 psi hydraulic we would get 750 x 14.52 or 10,890 psi leaving the nozzle and entering the mold. This force per area (pounds per square inch) acts on the projected area of the cavities to yield some amount of force in pounds trying to separate the mold halves. The clamp tonnage works to keep mold closed. The projected area is only the 2D area of molded part surfaces in the X and Y directions only – do not count sidewall area in the Z direction (part depth) ... in this example the Z axis/direction is in line with the screw.

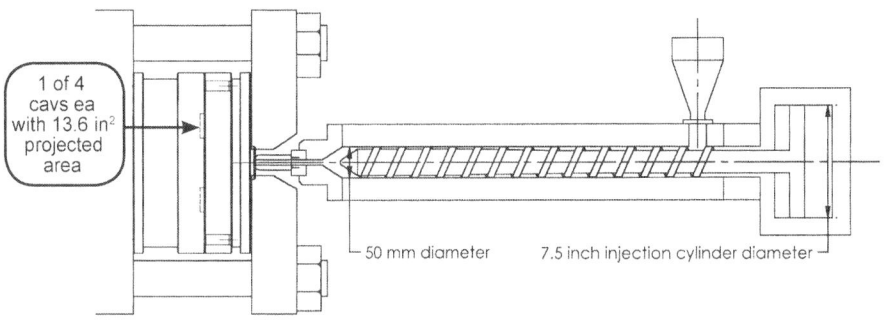

$$4 \ cavs \times 13.6 \ in^2 = 54.4 \ in^2 \ (mold \ proj. \ area)$$

$$\frac{10,890 \ lbs}{in^2} \times 54.4 \ in^2 = 592,416 \ lbs \ opening \ force$$

$$300 \ tons \times \frac{2000 \ lbs}{ton} = 600,000 \ lbs \ closing \ force$$

Note: This calculation for pressure amplification does ignore pressure drop which could be as much as 50%. The actual pressure varies at each location along the flow length – constantly being reduced by pressure drop in system. More sophisticated mold filling analysis software is needed to accurately quantify the average pressure in mold. If our equation shows ample tonnage working against full cavity pressure, we know we are conservative which insures success.

Injection Speed & Fill Rate

Using the same injection cylinder seen on previous pages, we can calculate the injection speed using the formula below (machine pump is given at 55 GPM):

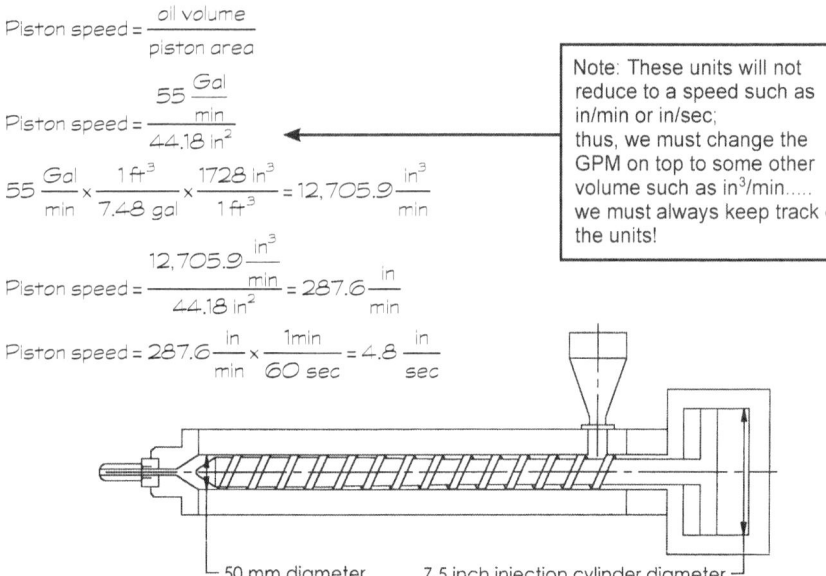

$$Piston\ speed = \frac{oil\ volume}{piston\ area}$$

$$Piston\ speed = \frac{55\ \dfrac{Gal}{min}}{44.18\ in^2}$$

$$55\ \frac{Gal}{min} \times \frac{1\ ft^3}{7.48\ gal} \times \frac{1728\ in^3}{1\ ft^3} = 12,705.9\ \frac{in^3}{min}$$

Note: These units will not reduce to a speed such as in/min or in/sec; thus, we must change the GPM on top to some other volume such as in³/min..... we must always keep track of the units!

$$Piston\ speed = \frac{12,705.9\ \dfrac{in^3}{min}}{44.18\ in^2} = 287.6\ \frac{in}{min}$$

$$Piston\ speed = 287.6\ \frac{in}{min} \times \frac{1min}{60\ sec} = 4.8\ \frac{in}{sec}$$

50 mm diameter   7.5 inch injection cylinder diameter

We calculate the volumetric injection rate (not injection speed) as follows:

$$50\ mm \times \frac{1\ in}{25.4\ mm} = 1.968\ inches\ (screw\ diameter)$$

$$area\ of\ a\ round = \frac{\pi\ d^2}{4}\ \ or\ \pi\ r^2\ (\pi = 3.1416)$$

$$\frac{3.1416 \times 1.968^2}{4} = 3.042\ in^2\ (area\ of\ screw)$$

$$area \times speed = volumetric\ injection\ rate$$

$$3.042\ in^2 \times 4.8\ \frac{in}{sec} = 14.60\ \frac{in^3}{sec}$$

$$14.60\ \frac{in^3}{sec} \times \frac{2.54\ cm}{1\ in} \times \frac{2.54\ cm}{1\ in} \times \frac{2.54\ cm}{1\ in} = 239.25\ \frac{cm^3}{sec}$$

If our part is polypropylene with a melt density of 0.70 gr/cm³ (0.90 cold density) and weighs 22.26 grams (same 4 cavity part described on previous page); we can calculate the fastest possible fill time:

$$4 \times 22.26\ grams = 89.04\ grams = shot\ weight$$

$$\frac{89.04\ grams}{0.70\ \dfrac{gr}{cm^3}} = 127.2\ cm^3 = volume\ of\ shot\ (melt)$$

$$\frac{127.2\ cm^3}{239.42\ \dfrac{cm^3}{sec}} = 0.53\ sec = fastest\ fill\ time\ possible$$

Cavity PSI Transducer Calculations

In order to plan or select the transducer, we need to calculate the full scale pressure. The molding press typically wants to know what 10 volts equals in terms of pressure (in 1st calc below: 10 v = 30,859.6 psi) This is important to know so we don't create a scenario whereby the full scale pressure is 10X (or even 50X) what the machine can generate. A typical molding press can create 10K - 40K psi in cavity. This writer has seen transducer applications whereby small pins are used with large charge amps whereby the full scale pressure is greater than 1MM psi; thus, resulting in poor resolution in output graphic displays ... this is corrected by using a smaller charge amp (with piezoelectric transducers) or a smaller max load transducer.

If using a piezoelectric type direct read pressure transducer (not behind an ejector pin). Determine transducer sensitivity value (pC/bar) and the charge amp full scale range (pC) ... use formula as follows:

$$\frac{\text{range (pC)}}{\text{sensitivity (pc/bar)}} = \frac{20{,}000 \text{ pC}}{\left(9.4 \dfrac{\text{pC}}{\text{bar}} \times \dfrac{1 \text{ bar}}{14.504 \text{ psi}} \right)} = 30859.6 \text{ psi}$$

in this example above, the charge amp is 20,000 pC and transducer is 9.4 pC/bar (piezo transducer) as shown.

If using a piezoelectric type transducer below an ejector pin, then the formula is altered as follows to convert the force back to a pressure by dividing by the ejector pin area:

$$\frac{\dfrac{\text{range (pC)}}{\text{sensitivity (pc/N)}}}{\text{area (In}^2)} = \frac{\dfrac{20{,}000 \text{ pC}}{\left(4.4 \dfrac{\text{pC}}{\text{N}} \times \dfrac{1 \text{ N}}{0.2248 \text{ lbs}} \right)}}{\left(\dfrac{\pi \times 0.0866 \text{ Inches}^2}{4} \right)} = \frac{173{,}480 \text{ lbs}}{\text{In}^2} = \text{psi}$$

If using a strain gage transducer, divide transducer load rating by ejector pin area (there are often different transducer load ratings available):

$$\frac{125 \text{ lbs}}{\dfrac{\pi \times 0.062^2}{4}} = \frac{125 \text{ lbs}}{0.003019 \text{ In}^2} = 41{,}403.47 \text{ psi}$$

The machine would send 10v to transducer and receive an output back based on load...if machine received 2.65 volts back, it would compute pressure as follows:

$$\frac{10 \text{ v}}{41{,}403.47 \text{ psi}} = \frac{2.65 \text{ v}}{X \text{ psi}}$$

$$\frac{2.65 \text{ v} \times 41{,}403.47 \text{ psi}}{10 \text{ v}} = 10{,}971.92 \text{ psi}$$

Miscellaneous Formulas

Thermal expansion:
$$\delta_t = \text{coeff} \times \Delta T \times \text{Length}$$

Cooling time:
$$Q = \frac{-(t^2)}{2\pi\alpha} \times LN\left[\frac{\pi}{4} \frac{(T_x - T_m)}{(T_c - T_m)}\right]$$

Hydraulic diameter:
$$D_h = \frac{4 \times \text{Area}}{\text{Perimeter}}$$

Reynolds number:
$$Nr = \frac{3160 \times \text{coolant flow (GPM)}}{\text{diam (inches)} \times \text{viscosity}}$$

Process potential:
$$P_p = \frac{\text{USL} - \text{LSL}}{6s} = \frac{\text{USL} - \text{LSL}}{6\sigma_{n-1}}$$

Process capability:
$$P_{pk} = \text{smaller of: } \frac{\overline{\overline{X}} - \text{LSL}}{3\sigma_{n-1}} \text{ or } \frac{\text{USL} - \overline{\overline{X}}}{3\sigma_{n-1}}$$

GPM required:
$$\frac{\text{Heat Load (Resin thruput X spec heat (resin) X } \Delta T)}{\text{spec heat (coolant) X temp rise of coolant}} = \text{lbs/hr of coolant reqd}$$

$$\frac{\frac{\text{lbs}}{\text{hr}} \text{of coolant required}}{\text{coolant} \frac{\text{lbs}}{\text{gal}} \text{X } 60 \frac{\text{min}}{\text{hr}}} = \text{GPM of coolant required}$$

Shrinkage:
$$\text{Shrinkage} = \frac{\text{cavity dimension} - \text{part length}}{\text{cavity dimension}}$$

$$\text{Cavity dimension} = \frac{\text{finished part length}}{(1 - \text{shrink rate})}$$

Thermal conductivity:
$$\text{Time (hrs)} = \frac{\text{Distance (ft)} \times \text{Heat (Btu)}}{\text{Therm cond (Btu/hr ft °F)} \times \text{Area (Ft}^2) \times \Delta T}$$

Trig (calc rise):

rise = Length × Tan ∠

or

rise = Length × 0.01745 × ∠

www.ingramcontent.com/pod-product-compliance
Lightning Source LLC
Chambersburg PA
CBHW072038190526
45165CB00018B/1080